Peterson
First Guide
to
Caterpillars
of North America

Amy Bartlett Wright

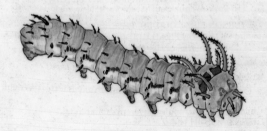

HOUGHTON MIFFLIN COMPANY

Boston   New York

Acknowledgments

For allowing the study of specimens and for their guidance in the preparation of illustrations and text, the author/illustrator thanks the following: David Furth and Edward Armstrong, Museum of Comparative Zoology; Charles Remington and Lawrence Gall, Peabody Museum of Natural History; William Winter; Donald Lafontaine; Gordon Pratt.

References

Covell, Charles V., Jr. 1984. *A Field Guide to the Moths of Eastern North America.* Boston: Houghton Mifflin Co.

Klots, A. B. 1951. *A Field Guide to the Butterflies of Eastern North America.* Boston: Houghton Mifflin Co.

Opler, Paul A. and Malikul, Vichai. 1992. *A Field Guide to Eastern Butterflies.* Boston: Houghton Mifflin Co.

Pyle, Robert M. 1988. *The Audubon Society Field Guide to North American Butterflies.* New York: Chanticleer Press.

Scott, James A. 1986. *The Butterflies of North America.* Stanford, California: Stanford University Press.

Stehr, Frederick W. 1987. *Immature Insects, Vol. 1.* Dubuque: Kendall/Hunt Publishing Company.

Tilden, James W. and Smith, Arthur C. 1986. *A Field Guide to Western Butterflies.* Boston: Houghton Mifflin Co.

Villiard, Paul. 1969. *Moths and How to Rear Them.* New York: Funk and Wagnalls.

Library of Congress Cataloging-in-Publication Data
Wright, Amy Bartlett.
    Peterson first guide to caterpillars / Amy Bartlett Wright.
    p. cm.
Includes bibliographical references and index.
ISBN 0-395-91184-2
1. Caterpillars—North America—Identification.
2. Caterpillars—North America—Classification.
I. Title. II. Title: Peterson first guides. Caterpillars.
QL548.W75   1993
595.78'04332'097—dc20   92-36585
                                                         CIP

Printed in Italy

NWI   15 14 13 12

Editor's Note

In 1934, my *Field Guide to the Birds* first saw the light of day. This book was designed so that live birds could be readily identified at a distance, by their patterns, shapes, and field marks, without resorting to the technical points specialists use to name species in the hand or in the specimen tray. The book introduced the "Peterson System," as it is now called, a visual system based on patternistic drawings with arrows to pinpoint the key field marks. The system is now used throughout the Peterson Field Guide series, which has grown to nearly 50 volumes on a wide range of subjects, from ferns to fishes, rocks to stars, animal tracks to edible plants.

Even though Peterson Field Guides are intended for the novice as well as the expert, there are still many beginners who would like something simpler to start with—a smaller guide that would give them confidence. It is for this audience—those who perhaps recognize a crow or a robin, buttercup or daisy, but little else—that the Peterson First Guides have been created. They offer a selection of the animals and plants you are most likely to see during your first forays afield. By narrowing the choices—and using the Peterson System—they make identification even simpler. First Guides make it easy to get started in the field, and easy to graduate to the full-fledged Peterson Field Guides. This one gives the beginner a start on a subject of great interest to gardeners, lepidopterists, and children of all ages: the caterpillars of North America.

The little caterpillars are so interesting; imagine living a good share of life in a form that many people mistake for a grub or worm, then going into a sarcophagus or coffin for a while, and finally emerging as a butterfly and dancing like an angel for a brief period before being recycled again.

Roger Tory Peterson

Introducing
the Caterpillars

Many people can identify several common butterflies. Not many, however, can put a name to more than a very few caterpillars. Even experienced naturalists are more familiar with adult butterflies or moths than with their immature form, the caterpillars. Yet these remarkable insects represent a vital stage in the life cycle of some of our planet's most beautiful and best-loved insects. They are also fascinating creatures in their own right, often astonishing or even frightening in appearance.

This book is the first popular field guide to caterpillars. It describes and illustrates 125 common species. The butterfly or moth into which each caterpillar will grow is included in the illustrations. The pupa, egg, and the caterpillar's preferred food plant are often shown in order to help identify the caterpillar.

The text identifies the caterpillars and adults by common names. These names may be regional, however, and a caterpillar may have a common name different from that of its adult—the Woolly Bear caterpillar develops to be the Isabella Tiger Moth, for example. For positive identification, you need the scientific, or Latin, name. The scientific name of each species is listed in the index.

Life Cycle Of Butterflies and Moths

The life cycle of a butterfly or moth consists of four stages of development: egg, caterpillar (or larva), pupa (or chrysalis), and adult. This process of changing from one form to another during growth is called metamorphosis. Flies, beetles, fleas, butterflies, and moths develop in this way. In the less advanced insects, such as cockroaches, bugs, and silverfish, the young are just smaller versions of the adults.

Egg. When the female butterfly or moth is ready to lay her eggs, she uses her senses of touch, sight, and smell to find the right spot. This is usually on or near the food plant, the plant that the caterpillar will feed upon as it grows. She attaches the egg to the chosen spot with a sticky fluid. Often the way a butterfly or moth places her eggs on the food plant is characteristic of her species; she might lay them in groups, singly, or in patterns. Eggs are usually laid during the warm weather of the growing season, but eggs of some species do overwinter—that is, they don't hatch until the next spring.

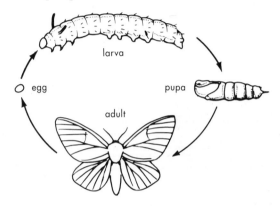

Caterpillar. An egg hatches at the time dictated by the caterpillar's genetic code, but hatching is also probably influenced by the environment. In spring, the days get longer, the weather warms, and food plants start their fresh new growth. Many species have more than one generation, or completed life cycle, in a growing season. In this case, the larvae from all generations will need fresh plant material to feed upon, so the signals for the larvae to emerge from the eggs can also occur in midseason.

After the minute larva, or caterpillar, cuts through its eggshell, it may devour the egg as its first meal, or it may start to feed upon the food plant. As it eats and grows, its body stretches until its skin becomes too tight. The larva then stops eating and molts, shedding its tight skin. Soon it is comfortable in a newly formed, looser skin and will resume feeding. Later it will outgrow that skin and molt again.

5

These stages between moltings are called *instars*; a caterpillar will go through several before it reaches full size in its final instar. This mature larva may bear little resemblance to its first instar.

Caterpillars are heavy feeders. In fact, their primary function is to eat and grow, and sometimes they do it very quickly. When there is an unlimited food supply and too few natural predators such as birds, caterpillar populations can soar. The Gypsy Moth's defoliation of vast areas of forests along the East Coast is an extreme example of the potential impact of just one species. Caterpillars can cause economic problems when they feed on cultivated plants and crops.

Pupa. If a caterpillar survives predators and disease long enough to grow to be a mature larva, it enters the next stage in the life cycle, the pupa. The pupa is the resting stage of the insect's life, during which it transforms into an adult. There are a few signs that indicate a mature larva is getting ready to pupate. The caterpillar might become lethargic and uninterested in eating. It may wag the forward portion of its body from side to side or wander away from the food plant. Many caterpillars change color just before pupating.

Once the caterpillar finds a suitable place, it transforms into a pupa, an immobile form protected from the environment. Many butterfly pupae are called chrysalises because of their gold or metallic spots (the word *chrysalis* has its root in the Greek word for gold, *khrusos*). Chrysalises usually hang upside down from a plant or are fixed upright to a twig. Pupating butterfly larvae do not make a cocoon around themselves, but pupating moth larvae may form a cocoon made of leaves, their own hairs, or silk that they spin. The cocoon is hidden in leaf litter or attached to a branch. Many other moth pupae have no outer protective case at all. Sphinx moth caterpillars, for example, dig underground and become hardened pupae without cocoons, sheltered at or just below the soil surface. Pupae come in many shapes and sizes—each species looks different.

Adult. To watch a beautiful and graceful flying creature emerge from its immobile pupa is to witness a fascinating and seemingly miraculous birth.

Butterflies push and break open a spot on the hanging chrysalis until it splits far enough for the butterfly to pull its body out. The emerged adult hangs from the shed chrysalis or a twig until it is ready to fly. Moths also push open a hole in the pupa and cocoon. Emerging moths may need to climb out of the soil or the leaf litter to reach a twig.

Minutes after emerging, the adult butterfly or moth holds out its limp wings and flaps them slowly. This movement pumps blood into the wing veins, expanding the wings to their full span. Within a short time, the wings stiffen. The adult is ready to fly, perhaps alight on flowers, drink nectar or water, and search for a mate. With mating, the life cycle is again begun with a new generation.

Structure of a Caterpillar

Although there is much variety among the 11,000 species and about 80 families of butterflies and moths occurring in North America north of Mexico, caterpillars share a basic body structure.

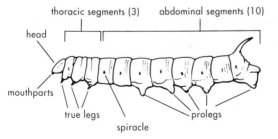

The caterpillar head has simple eyes, two short **antennae,** and strong chewing **mouthparts.** The soft body is made up of 13 segments. The first three segments behind the head are called the **thoracic segments.** Each of these three segments has a pair of jointed **true legs.** These will become the legs of the

7

adult butterfly or moth. The remaining 10 body segments are the **abdominal segments.** These usually contain 5 pairs of false legs called **prolegs**. Their function is purely to support and move the growing mass of the body as the caterpillar eats its way to maturity. Loopers have no prolegs on some of the middle segments, necessitating the looping or "inchworm" method of moving. Small hooks called crochets at the bottom of the prolegs act as grippers for the caterpillar, allowing it to hold tightly onto a branch or leaf.

Spiracles are circular openings on each side of the caterpillar body through which air passes in and out of the body. All caterpillars have them, though in some species they are hard to see.

Protection and Defense

Caterpillars are exposed and vulnerable when feeding. In order to survive, many physical and behavioral adaptations have evolved that protect larvae from birds and other predators. Some caterpillars have alarming-looking structures such as the fierce red horns of the Hickory Horned Devil (page 58). Most swallow-tail larvae are equipped with a protrudable forked scent gland called an *osmeterium,* which emits a foul odor. Other caterpillars are covered with spines containing painful toxins that birds—and collectors—quickly learn to avoid.

Some larvae build structures in which they hide from predators. Bagworms (page 124) live in silk cases camouflaged with plant material, and some skipper larvae cover themselves with leaf tents. The Scalloped Sack-bearer (page 124) larva builds a portable "house" that it can carry about in search of new leaves.

The coloration of a caterpillar can also help it survive. Bright "eyespots" on the Spicebush Swallowtail larva (page 16) may startle away an approaching bird. Some caterpillars escape detection because their coloration mimics something else. The young Red-spotted Purple larva (page 48), for example, looks like a bird dropping. The colors of some larvae, such as the underwings, serve as camouflage by blending in with the immediate surroundings.

Other larvae combine color and posture for survival. Faced with an intruder, the Black-etched Prominent (page 74) lifts up its head and rear end threateningly, exposing a seemingly huge face and long pink rear tentacles. The Large Maple Spanworm (page 32), camouflaged in the colors of tree bark, clasps a branch with its prolegs and stretches out taut and still; it looks remarkably like a little twig.

Raising Caterpillars

Raising caterpillars by providing them with food and shelter until they transform into adults is enjoyable and educational. There is still much to be discovered about the immature stages of butterflies and moths, and rearing them provides an opportunity for close observation.

You can begin with eggs found on the food plant (food plants are named for each species described in this book). It is also possible to collect eggs from an adult female moth. Usually the female's abdomen is rounder and fuller than the male's. An easy way to find female moths is to net them near a light at night. You must identify the adult so you can provide the caterpillar with food it can eat. Place the moth unharmed inside a paper bag, and she may lay eggs inside the closed bag. Later, release the moth unharmed and find the eggs she fixed to the paper. Remove the eggs by cutting out the paper around them, and place the paper pieces with eggs on them in a jar or other container with the known food plant. As the newly hatched caterpillars grow, they will need plenty of food and space. To provide fresh air, punch holes (not big enough for the larva to escape through) in a large jar lid, or cover the jar opening with a piece of gauze or a fine screen mesh held on with a rubber band. Put in some twigs or narrow branches for the caterpillar to climb on when it begins to pupate. If your caterpillars are of a species that pupates below ground, provide them with light loose soil at least one inch deep on the bottom of their container. Emerging adults need room to climb up and spread their wings before flying.

If you prefer to start with caterpillars rather than eggs, collect them from plant leaves or tree branches. You can sometimes knock larvae from tree or bush branches with a stick or pole. Place a white sheet on the ground below so none escapes notice.

Most important, watch each stage carefully. Observe how the egg changes color before hatching, how the caterpillar eats and moves, and how it changes features and perhaps colors as it grows. Be sure to look at your pupa often so you can release the adult butterfly or moth outdoors soon after it emerges.

Butterflies and Moths: the Adults

Unless you can recognize a caterpillar or its family, it is difficult to know whether it will become a butterfly or moth.

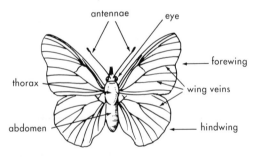

Butterfly and moth adults have similar basic anatomy. They have a head with an **eye** on each side and two **antennae** on top. Most have a coiled hollow tongue or **proboscis** for sipping nectar or water. The **thorax** bears the 3 pairs of legs below and 2 pairs of wings. The **abdomen** contains the internal organs.

There are two pairs of wings, the **forewings** and the **hindwings.** The two pairs overlap, with the forewings above the hindwings. When a moth's wings are folded, its hindwings may not be visible. A butterfly at rest or feeding may fold its wings upright over its body, so that only the underside of the hindwing is exposed. Many butterflies rest or feed in this posture.

The body and the wings are covered with overlapping scales. The colors of the scales give the adult its color and markings. The wings consist of membranes that are supported by **veins**. Each butterfly or moth family has a characteristic pattern of wing veins.

How can you tell whether an adult is a butterfly or a moth? In general, butterflies usually have knobbed antennae, they usually fly during the day, and they hold their wings vertically above their bodies when at rest. Moths have simple or feathered antennae, they hold their wings flat or rooflike over their bodies when resting, and they usually fly at night or in the evening.

Attracting Butterflies

In order to get butterflies to lay their eggs in your yard, simply grow the plants the caterpillars like to eat. For example, if you grow parsley, you may attract Black Swallowtails. The females will lay eggs on the leaves, and you can watch the eggs hatch and the caterpillars grow. In this guide, food plants are named for each of the larvae. *Keep your lawn and garden chemical free.*

You should also grow the plants that the adults feed on. Butterflies drink nectar, the sweet juices of flowers, and certain flowers will bring the adults around. Plants with varying blooming cycles can be placed together to keep the garden full of activity throughout the growing season. Adults respond to flower shape, color, and scent. There are many books on butterfly gardening available.

How To Use This Book

This simplified field guide presents 125 common caterpillars found in various areas throughout the continental United States. The caterpillars are grouped by their appearance.

To identify a caterpillar, first study it closely. Does it have horns or a "tail"? Bristles or bumps? Is it smooth or hairy? Compare your caterpillar's most obvious physical character-

11

istics with those of the 11 groups described and illustrated below. Match your caterpillar with the general heading that best describes it and turn to the appropriate pages. The illustrations and text will give you information on coloration, markings, habits, food plants, habitat, and geographic range to help you pinpoint the identity of your caterpillar.

A few caterpillars have irritating hairs or spines that are painful to touch. These caterpillars are identified with this symbol:

Smooth   **Pages 16–47**

The bodies of smooth caterpillars are generally hairless, with no protuberances.

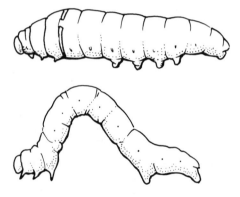

Smooth With Knobs   **Pages 46–59**
Or Bumps

These caterpillars are generally hairless, but their bodies have some protuberances, often with single hairs.

Smooth With Rear Horn Or "Tail" Pages 60–69

Caterpillars of this group are generally hairless, and the last abdominal segment has a single pointed projection. The tail may be just a stump.

Smooth With Fleshy Filaments Pages 70–75

These caterpillars have generally hairless bodies, with fleshy filamentlike projections.

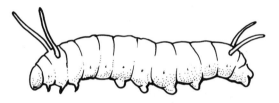

Sluglike Pages 76–81

The body shape of this group is short and wide; legs are not visible. They are smallish caterpillars, and may be hairless or downy.

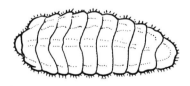

Hairy   **Pages 82–93**

These caterpillars' bodies are covered with hairs of more or less the same length.

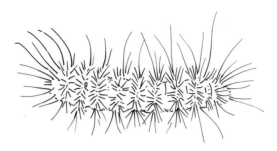

Hairy With Tufts   **Pages 92–97**
Or "Pencils"

The bodies of these caterpillars are covered with hairs. Some tufts or "pencils" stand out.

Bristled   **Pages 98–101**

Caterpillars of this group have bodies covered with short, stiff hairs.

Branched Spines   **Pages 102–121**

Thick spines with sharp narrow branches cover the bodies of these caterpillars.

Internal Feeders   **Pages 122–123**

Look for these caterpillars inside the stems and vines of plants where they feed. They are very pale in color.

Structure-building   **Pages 124–125**
Caterpillars

These caterpillars build protective structures around themselves.

Smooth

TIGER SWALLOWTAIL   To 2 inches

The newly hatched caterpillar is brown and white and looks like a bird dropping. As it matures, however, the larva becomes smooth and green, enlarged in the front with *2 orange eyespots* and a *yellow band.* The caterpillar may be hard to find because it usually feeds high in the treetops, making a shelter by folding the edges of a leaf together. It feeds on tree leaves, including willow, cherry, tulip tree, poplar, basswood, and birch. Just before pupating, the larva may turn brown. The chrysalis is suspended from a twig or other support for overwintering. The egg is round and yellow-green.

The adult is one of our most familiar butterflies. It is a very large (up to 6 inches), high-flying insect, yellow with black tiger stripes. The female has a dark form in which the yellow is mostly replaced by black. The long "tails" on its hindwings give the family its name. The Tiger Swallowtail is found in the eastern half of the United States including most of the Great Plains. The similar Western Tiger Swallowtail is found west of the Rockies.

SPICEBUSH SWALLOWTAIL   To 1 1/2 inches

The larva feeds on spicebush and sassafras, plants that are found in open woods, woodland borders, and meadows, and beside streams. The caterpillar is dark green, with two large *black and orange eyespots* and four lines of small *blue spots* in the middle of its back. An older larva lives in a shelter made by curling the edges of a leaf upward and attaching them with a silken mat. It overwinters as a chrysalis; the egg is round and pale green.

The large (up to 4 1/2 inches) Spicebush Swallowtail butterfly is a mimic of the Pipevine Swallowtail (page 74), which is bad-tasting to birds. The Spicebush Swallowtail is generally an eastern butterfly, seen from Canada to Florida and Texas, becoming less common west of the Mississippi River.

TIGER
SWALLOWTAIL

SPICEBUSH
SWALLOWTAIL

pupa

17

Smooth

BLACK SWALLOWTAIL   To 2 inches

This caterpillar is also called **Parsley Caterpillar.** The young larva is black with a white saddle, but when mature it is smooth and green with *black bands* and *yellow spots*. It is well concealed in the leaves of its host plant. It has an orange osmeterium (see page 8). The Parsley Caterpillar is bad-tasting to predators such as birds, because it absorbs the toxins of the plants it eats. Its food plants are members of the carrot family: Queen Anne's lace, carrot, parsley, and dill. It overwinters as a chrysalis.

The adult Black Swallowtail has a wing-span of up to 4½ inches. Its coloration, black with spots along the edges of its wings, mimics that of the bad-tasting Pipe-vine Swallowtail, which may give it some protection from predators. It is similar to Baird's Swallowtail of the West. Eggs are round, yellow to cream. It is found in southern Canada and throughout the eastern United States, as well as in the southwestern states and Mexico.

GIANT SWALLOWTAIL   To 2½ inches

The caterpillar is called **Orange Dog** because it feeds on citrus plants and is sometimes considered a pest to citrus growers in the Southeast. The young caterpillar is brown and white, and looks very much *like a bird dropping*. The mature larva is brown with creamy markings including a *white saddle* in the middle. Its *orange-red* osmeterium, or scent gland, emits a foul odor when the caterpillar is disturbed. Most swallowtail larvae have this organ for defense. The yellow-green eggs are laid singly on host plants. This species passes the winter as a chrysalis.

The Giant Swallowtail, with a wingspan of up to 6 inches, and the female Tiger Swallowtail are the largest butterflies in North America. The Giant Swallowtail occurs throughout much of the United States.

BLACK
SWALLOWTAIL

GIANT
SWALLOWTAIL

ORANGE DOG

Smooth

ZEBRA SWALLOWTAIL To 2 inches

The larva is green with *one bold black band* and *many yellow ones* crossing its back. A dark form of the larva is mostly black with white and yellow bands crossing its back. It eats pawpaw and its plant relatives. It is found in wet areas, shrubby borders, marshes, and wooded riversides where the host plants thrive.

The adult Zebra is one of the North American kite swallowtails, a tropical group named for the triangular shape of their wings and the long "tails" or extensions of the hindwings. This species is named for its black and whitish "zebra stripes." Spring-emerging adults are smaller and paler with shorter tails than those emerging in the warm summer months. Zebra Swallowtails have a wingspan of about 3 inches. The round, pale green eggs are laid singly. Winter is passed in the pupal stage. This butterfly is found from southern New England west to the Great Plains and south to Florida and the Gulf States.

CLODIUS PARNASSIAN To 1¼ inches

The larva is *black* with short black hair and *orange, yellow, or red spots*. It feeds at night on bleeding heart and is found in meadows and rock outcroppings. Like its relatives the swallowtails, it has an osmeterium but rarely uses it. It absorbs toxins from its food plant and is poison to predators.

The adult butterfly has white wings, often with transparent areas, and black antennae, and measures 2–3 inches across. All mated female parnassians have a waxy patch on the tip of the abdomen, placed there by the male, which prevents further mating. The range of the Clodius Parnassian is limited to North America, primarily the Pacific Northwest into Alaska, Montana, Idaho, Utah, and Wyoming.

ZEBRA
SWALLOWTAIL

CLODIUS
PARNASSIAN

Smooth

CLOUDED SULPHUR and ALFALFA BUTTERFLY
To 1 inch

The larvae of these two butterflies are nearly identical. They are *bright green* caterpillars, darker on the back and with *light side stripes.* They eat a variety of legumes; the Clouded Sulphur prefers clover, the Alfalfa caterpillar favors alfalfa. They are very common in open fields and meadows.

The butterflies, which measure about 2 inches in wingspan, are among the most commonly seen across the United States, except in areas of thick forests or deserts. Winter is passed by the green chrysalis.

CHECKERED WHITE
To $1\frac{1}{8}$ inches

This caterpillar is *blue-green* with black speckles and *2 yellow stripes* along each side. It has a light downy hair covering and feeds upon wild and cultivated crucifers such as mustards and cabbage. The adult butterfly is seen early in the spring and through the summer across the United States but is more common in the Southwest. It is white with regular darker markings and measures about 2 inches across. The chrysalis overwinters.

IMPORTED CABBAGEWORM
To 1 inch

This caterpillar is very common in gardens and fields where cabbages and their relatives grow. The larva is blue-green with a *yellow line* along its back and a covering of very short hairs. It is usually seen eating on the underside of the leaves but can bore into the cabbage head.

The adult, the common **Cabbage Butterfly,** is mostly white with black tips on the forewings, and has a wingspan of 2 inches. There are round black spots on both wings. The butterfly was introduced from Europe around 1860 and now ranges across North America. It may have 2–8 broods in a growing season, with the last generation of the year overwintering as a chrysalis.

ALFALFA BUTTERFLY

CLOUDED SULPHUR

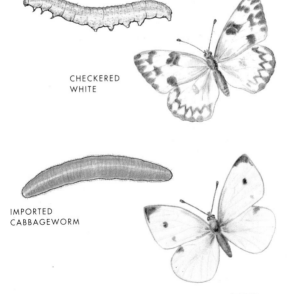

CHECKERED
WHITE

IMPORTED
CABBAGEWORM

CABBAGE BUTTERFLY 23

Smooth

CLOUDLESS SULPHUR   To 1 inch

The larva is green with *yellow side stripes*. It may turn orange just before pupating. It eats wild senna.

The Cloudless Sulphur butterfly is one of the group called the giant sulphurs; it can measure up to 3 inches across its yellow wings (large for a sulphur). This migratory butterfly spends the winter in Florida and the Gulf States. Warmer weather finds it as far north as the Canadian border and west to southern California.

SLEEPY ORANGE   To 1 inch

The larva is green to *grayish green* and downy, with yellow side stripes. It eats legumes such as senna and clovers, tip first. Its pupa is green to black.

The butterfly is small and orange with broad black borders on wings that span 2 inches. The adult overwinters in the South. It is very common in the South and Southwest.

AMERICAN SNOUT   To 1 inch

This is a dark green caterpillar with *yellow side stripes* and a *pale green underside*. It eats hackberries, preferring young leaves. The pupa is green.

The American Snout, with a wingspan of about 2 inches, belongs to a small family of mostly tropical butterflies. The "snout" on the front of the butterfly's head is really lengthened mouthparts. The snout is part of the butterfly's camouflage; with its wings folded upright, the butterfly looks like a leaf, and the long mouthparts resemble the leaf's stem. This species is found in the southern United States, where adults overwinter. Summer populations migrate northward in the spring.

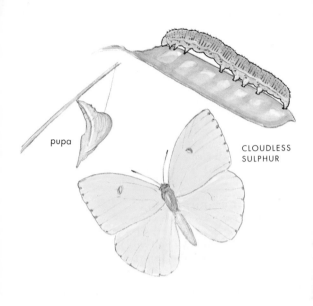

pupa

CLOUDLESS
SULPHUR

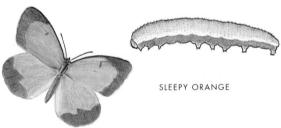

SLEEPY ORANGE

AMERICAN
SNOUT

25

Smooth (With Forked Rear End)

TAWNY EMPEROR To 1⅓ inches

This is a mostly green caterpillar. It has 2 *wide yellow stripes* on its upper sides and straight yellow bands on its lower sides. The last segment appears to be *split* into 2 ends; the head has 2 protuberances with *short branches.* It feeds upon hackberry. The pupa usually lies flat in a silk mat on the underside of a leaf. This species is found in the eastern United States south of northern New England. The adult butterfly has a wingspan of about 2½ inches.

HACKBERRY BUTTERFLY To 1¼ inches

This caterpillar is very similar to that of the Tawny Emperor, but the Hackberry caterpillar has *yellowish V markings* along its sides, and its *yellow bands* on the upper sides are narrower. The *branched knobs* on the head are generally longer than the Tawny's. It eats hackberry, especially young leaves, in deciduous wooded areas. The light green eggs are usually laid singly on new growth.

This 2-inch butterfly is lighter in color than the Tawny Emperor and is more common. It is seen in the eastern United States south of northern New England to the Dakotas and south into Mexico.

RINGLET To 1 inch

The caterpillar is green and covered with small white bumps. Along each side it has a pale green stripe above a *yellow-white stripe.* The short *forked end* is *pink,* and the head is green. Its colors can vary from green to brown, camouflaging it in open grassy areas. It feeds on grasses. The pupa is green or brown with 9 black stripes.

The adult Ringlet is a small butterfly with yellow or orange-brown wings. It usually has a round black eyespot on the underside of each forewing. Its range is variable, but it can be found in most of the northern United States. In the West it occurs widely from Alaska to Baja California. Its range is still expanding.

TAWNY EMPEROR

pupa

HACKBERRY BUTTERFLY

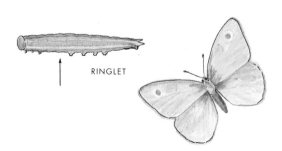

RINGLET

Smooth (With Forked Rear End)

COMMON WOOD NYMPH   To 1 $^1/_4$ inch

The caterpillar is green with *4 lighter yellow stripes* and is covered with very short hair. The final segment appears to be split into *2 reddish points.* It feeds on grasses and over-winters as a first-instar larva. The pupa is green, found in woods and meadows and fields throughout much of North America.

The adult is the only wood nymph butter-fly in the East. The female, at up to 3 inches across, is usually larger than the male, lighter in color, and with larger eyespots. This butterfly's flight is quick and erratic, making it hard to catch.

COMMON ALPINE   To 1 $^1/_2$ inches

The caterpillar is green with *yellow side stripes;* the *yellow-brown* head is *large.* It is believed to feed on grasses and sedges in high mountain meadows. The body is tapered, ending in *2 points.* The partially grown larva overwinters. The mature larva pupates among leaves stuck together with the caterpillar's silk. The pupa is whitish brown with brown stripes.

The Common Alpine adult, about 2 inches across, usually has eyespots ringed with yellow or orange on both pairs of wings. It is a common butterfly in mountains and just into lower elevations in the Rocky Moun-tains, north into Canada and Alaska.

NORTHERN PEARLY EYE   To 1 $^1/_2$ inches

The caterpillar is green with *dark and yellow stripes.* The head has 2 unbranched *reddish horns.* The last segment is *forked* and *red* at the tips. It eats grasses in forests, so look for it in wooded areas and their edges. It over-winters as an immature larva. The pupa is pale green or blue-green.

The adult butterfly measures up to 2 $^1/_2$ inches across and is found in the north and central areas of the United States. The adult might be seen resting on tree trunks, or feeding on sap or bird droppings.

top view
of "tails"

COMMON
WOOD NYMPH

COMMON ALPINE

NORTHERN
PEARLY EYE

29

Smooth

CABBAGE LOOPER   To 1 1/4 inches

This caterpillar is considered a pest because it eats a wide variety of garden crops, such as cabbage, corn, watermelon, tobacco, and herbs. It is a green caterpillar, darker above than below, and walks with a *"looping" or inching gait* because there are *no prolegs on 2 of the middle segments.* As the rear end walks forward, the body bends up in the middle, making a loop. Then the front of the body reaches up and the true legs grasp forward as the body straightens out.

The adult is a small, brown moth with a wingspan of 1 1/2 inches; notice the white curved marking near the center of the dark forewing. Many generations grow in a year. It is common throughout the United States.

CELERY LOOPER   To 1 1/4 inches

This is very similar to the Cabbage Looper above. It feeds upon a variety of plants also, including clover, lettuce, corn, plantain, toadflax and other plants. It also *moves by "looping."* Here it is shown at rest, with its body straight.

The Celery Looper Moth is small and common across the United States. It is active day and night.

CURVE-LINED ACONTIA   To 1 1/4 inches

This larva also is a *looper*, with *3 pairs of prolegs* rather than 5. Its coloration *mimics that of a bird dropping,* allowing the caterpillar to escape the attention of predators. As a young larva it is mostly black. Its food plants include swamp rose-mallow, found in brackish or saltwater marshes near the coast. The illustration shows the loop it makes when walking.

The small adult moth has forewings marked with red-brown patches and yellow bands. Its range includes the eastern central United States.

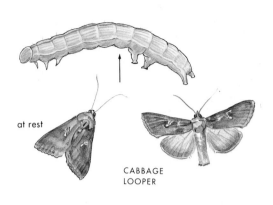

at rest

CABBAGE
LOOPER

CELERY LOOPER

CURVE-LINED ACONTIA

Smooth

LARGE MAPLE SPANWORM   To 1¹/₂ inches

This larva has only *2 pairs* of prolegs and walks by *looping* (see page 30). It is gray-brown and *mimics a twig* in both color and posture. It has many food plants but favors maple and oak. It overwinters as a pupa.

The Large Maple Spanworm Moth, with a 2-inch wingspan, is big for a member of the inchworm family. This common moth occurs across the United States.

FALL CANKERWORM   To 1 inch

This larva has *3 pairs* of prolegs rather than 5. It feeds on many kinds of trees and can defoliate elms and apples. There are two color variations, a light green-brown form with *lighter side stripes* and a dark form with reddish brown sides, also with lighter side stripes. The larva can lower itself from a tree by hanging from long strands of its own silk. The wind can then blow it to a new feeding place. The mature larva pupates in the soil through the summer. In autumn, the wingless Fall Cankerworm Moth female lays her eggs in a compact mass around small twigs. The eggs overwinter. The male has a 1-inch wingspan. This species ranges from southern Canada throughout most of the United States.

LINDEN LOOPER   To 1¹/₂ inches

This inchworm has *2 pairs* of prolegs. The *band* along its back may be dark or yellow-orange. It feeds upon linden and other deciduous trees, often defoliating them.

The Linden Looper Moth is also called the **Winter Moth.** The female adult is wingless; the male measures at most 1¹/₂ inches. It is found in the northern half of the United States west to the Rockies. The larva pupates in early summer just below the soil surface. The moths emerge in late fall and lay eggs that overwinter. There is one generation a year.

LARGE MAPLE
SPANWORM

FALL
CANKERWORM

male

LINDEN LOOPER

33

Smooth

SILVER-SPOTTED SKIPPER   To 2 inches

This is a common skipper in suburban areas, meadows, wastelands, and watersheds where its preferred food plant, locust, grows. It also eats other legumes. Skipper caterpillars are notable for having a *large head* and a *visible "neck."* This larva's head is *large*, black or brownish red, and gains *two orange spots* at maturity. The body is green to yellow-green, often with narrow, *vertical*, darker stripes. The key to finding the caterpillar is to look for the leaf shelter it makes on the host plant. A young larva makes this small shelter by cutting out a circular flap from a leaf and folding it over its body. An older larva makes a larger shelter of two leaves held together by silk threads. The pupa overwinters.

Many scientists consider skippers separate from butterflies and moths because they have characteristics of both groups. Notice that the antennae are thickened and curved at the tips. The Silver-spotted Skipper is identified by the silver patches on the underside of its hindwings. It has a wingspan of $2\frac{1}{2}$ inches. Like most skippers, it is a strong, fast flier with jerky, or "skipping," movements. It is commonly seen on flowers in suburban backyards. Its range, throughout the continental United States and Central America, is among the broadest of our butterflies.

LONG-TAILED SKIPPER   To 1 inch

Because the larva eats cultivated beans, it is called the **Bean Leaf Roller.** It is yellowish green or gray with a yellow line along its back and a *yellow and green line* along each side. It is covered with *many black and yellow dots.* The head is *large* and *brown.*

The adult Long-tailed Skipper has "tails," or extensions, on its hindwings like a swallowtail. It measures up to 2 inches across. At least 3 generations grow each year. It is common in the Southeast but also migrates north in spring.

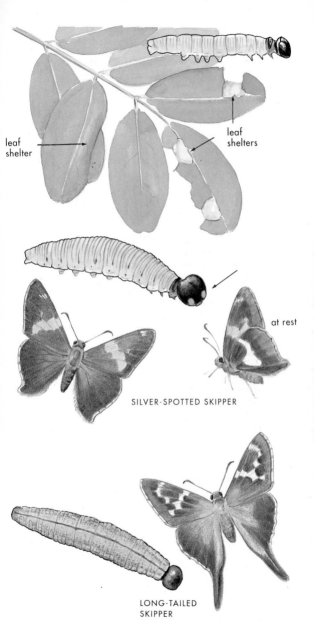

leaf shelter

leaf shelters

at rest

SILVER-SPOTTED SKIPPER

LONG-TAILED SKIPPER

35

Smooth

COMMON SOOTY WING   Less than 1 inch
The larva of this common skipper is green with a downy hair covering. It makes a shelter in the rolled leaves of its host plant, pigweed, or lamb's-quarters. When full grown, it hibernates inside its shelter. The pupa is green to brown with some hair.

   The small adult skipper is also appropriately called Roadside Rambler; its flight is quick and low to the ground. It is common throughout North America.

DREAMY DUSKY WING   Less than 1 inch
The larva is mostly green with *small whitish dots,* a dark line on its back, and light side stripes. The head is *red-brown.* It eats aspen, willow, poplar, and other plants and makes a nest of leaves rolled or tied together with silk. The mature caterpillar overwinters, and the pupa is brown.

   The adult, a small skipper, flies in early spring in most of the United States except the South and Great Plains.

FUNEREAL DUSKY WING   Less than 1 inch
This larva feeds on alfalfa, clover, and other legumes. It is yellow-green with a *black head.* It has a *dark green line* on its back and *yellow* side stripes. The larva overwinters.

   The adult is a western skipper with white fringes on its hindwings. It ranges from western Kansas to California and Mexico.

FIERY SKIPPER   To $^3/_4$ inch
This larva is pale green, yellow-brown, gray, or greenish brown. It has a *brown stripe* along its back and each side and a *dark collar* behind the *black head.* It eats grasses, especially crabgrass and Bermuda grass.

   The adult is a small skipper with yellow wings; the female is darker than the male. Its antennae are short. The Fiery Skipper is abundant in the South and ranges north to Connecticut, Michigan, and Nebraska.

COMMON
SOOTY WING

shelter

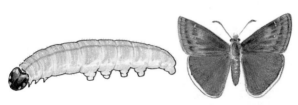

DREAMY DUSKY WING

FUNEREAL DUSKY WING

FIERY SKIPPER

at rest

37

Smooth

VARIEGATED CUTWORM   To 1¹/₂ inches

The caterpillar is gray or light brown with
darker mottling. It has a *light yellow dot* on
the back of most body segments and a *dark
W* near the tail end. Cutworms feed on
young plants, cutting the stems off at soil
level and causing a great deal of damage to
gardens and crops. This species can also
climb up the plant and feed. The larva can
often be seen just under the soil surface in
its *"C" posture.* Pupation occurs in the soil.

Adult cutworm moths have dark forewings and light hindwings. This species
ranges throughout North America.

ARMYWORM   To 2 inches

The armyworms are cutworms that move in
masses to fresh feeding areas. In a population peak, they can destroy vegetation over
vast areas. The dark green or brown larva
has *dark stripes* along the back and sides. It
feeds on corn, legumes, and other crops.
The partially grown larva overwinters under
the soil, pupating after feeding in spring.

The small, pale gray or brown Armyworm
Moth is active at night. It is found east of the
Rockies and elsewhere around the world.

YELLOW-STRIPED   To 1¹/₂ inches
ARMYWORM

This larva has two color forms. The light
form (shown) has yellow lines along its sides
and 2 rows of *black triangles* along its back.
Its head is dark with a *white inverted Y* in
front. The dark form is mostly black with a
yellow stripe along each side. This caterpillar is also called the Cotton Cutworm
because it feeds on cotton bolls. It is a
common pest in crops in the Southwest and
the eastern half of the United States. There
are many generations a season; the pupa
overwinters.

The adult has white markings on brown
forewings. Like other armyworm and cutworm moths, the underwings are pale and
unmarked, with or without dark borders.

"C" position

VARIEGATED
CUTWORM

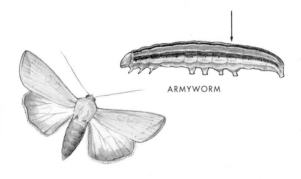

ARMYWORM

YELLOW-STRIPED
ARMYWORM

pupa

Smooth

CORN EARWORM   To 2 inches

This major agricultural pest is the caterpillar most often found in fresh ears of corn. It
also feeds heavily on cotton and the fruits of
tomato plants (another name for this larva is
the Tomato Fruitworm). The caterpillars are
cannibalistic—they will eat each other. The
Corn Earworm can be green, pink, or brown
in color. The mature larva drops from the
fruit and burrows under the soil to pupate
over the winter. Small round eggs are laid
singly on the host plant.

The yellowish tan adult moth has a wingspan of less than 2 inches. It is common
across the country but overwinters in the
South, moving north as the weather warms.
It is also found around the world.

ZEBRA CATERPILLAR   To 1³/₄ inches

The larva has a *black center band* and is
yellow on each side with *dark vertical markings* within the yellow. The head is orange to
reddish. It prefers cabbage and other crucif-
erous plants but also feeds on a variety of
other garden and field crops. It overwinters
as a pupa.

The Zebra Caterpillar Moth has dark forewings with light markings and yellowish
white hindwings. There are usually two
broods a year. It ranges from southern
Canada south through Virginia and west to
California.

BROWN-HOODED OWLET   To 1¹/₂ inches

This larva feeds on the flowers of low-growing plants, preferring asters and goldenrod.
The sides are yellow with many *black vertical bars*, and it is striped lengthwise in red
and black along its back. The shield behind
its head is *black*, and there is a *black hump*
on a segment near the rear.

The adult moth has a line of reddish
brown tufts along its back. It is found from
Nova Scotia to South Carolina, west to California and Saskatchewan.

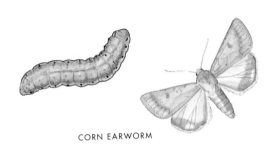

CORN EARWORM

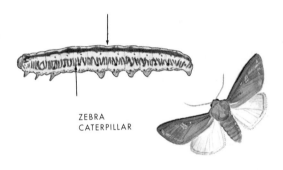

ZEBRA
CATERPILLAR

BROWN-HOODED
OWLET

41

Smooth

IRENE UNDERWING
MOTHER UNDERWING
To 3$^1/_2$ inches

The larvae of these two species are nearly identical and not easily distinguished from each other. The illustration shows one in side view and one seen from above. Like many underwing larvae, they are colored very much like *tree bark* and have a *raised area* on the eighth segment. The Irene Underwing feeds on poplar and willow; the Mother Underwing eats willow. Young larvae seem to hug the twig with their bodies; older larvae walk or loop. All of the underwings pass the winter in the egg stage.

During the day, the large adult moths rest with their wings folded, hiding the colorful underwings that give this group its name. Their forewings resemble tree bark, so they are well camouflaged until disturbed to flight. Then the bright underwings are displayed, showing a flash of color that may startle a hungry bird long enough to let the moth escape.

Adults of these two species can be told apart by the patterns of the forewings. Their hindwing colors differ, too: The areas between the black bands on the hindwing of the Irene Underwing are red, while those of the Mother Underwing are orange. These moths have a wingspan of about 3 inches. The Irene Underwing can be found in California and states immediately east. The Mother Underwing ranges from Nova Scotia south to Maryland and west to South Dakota and Missouri. It can be common, but it is rarely seen in the South.

IRENE
UNDERWING

MOTHER
UNDERWING

pupa

cocoon

43

Smooth

MAGDALEN UNDERWING   To 3 inches
This underwing larva is colorfully marked with *black vertical bands* and *orange spots* at the sides and yellow above. Its food plant is honey locust. As with other true underwings, the eggs overwinter.

The hindwing of the adult moth is yellow between the black bands. The range of this underwing is in the central and parts of the eastern United States. It is more commonly seen in the western areas of its range.

EIGHT-SPOTTED   To 1¹/₂ inches
FORESTER
This caterpillar has vertical *black and white bands* with one *black-spotted orange band* on each segment. The head is orange with black spots. It has little hair. Its food plants include grape, Virginia creeper, and Boston ivy. It is found at the edge of woods, in vineyards, and in cities. There are 2 generations a season in the South.

The adult moth has black body and wings, with a total of 8 spots: 2 yellow spots on each forewing and 2 white spots on each hindwing. The day-flying Eight-spotted Forester is found in the eastern half of the United States.

TOADFLAX CATERPILLAR   To 1¹/₂ inch
The Toadflax Caterpillar is *bluish* with *yellow stripes;* 2 rows of *black bars* on its back and *black spots* on each side. It is native to central Europe and was introduced to Canada in the 1960s in an effort to control toadflax, a common pasture weed and the larva's only food plant. A population became established in Southern Ontario and has spread east and south to Pennsylvania. It pupates in a cocoon that includes bits of the food plant.

The small adult moth has 8 dark stripes along the outer forewing, giving the wings a fringed look.

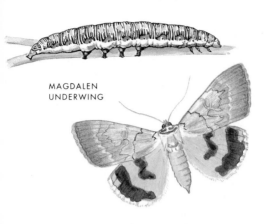

MAGDALEN
UNDERWING

EIGHT-SPOTTED
FORESTER

TOADFLAX CATERPILLAR

cocoon

Smooth

WHITE UNDERWING   To 4 inches

The caterpillar is *whitish* with *2 dark saddles*. The eighth segment is *slightly humped*. The larva has white *fringes* along the undersides of its body that blur the edges of its shadow, making it less visible to its enemies. The top of the head has orange markings. This underwing larva is most common on aspen but also eats poplars and willows. It overwinters as an egg.

The adult moth is also called the Relict. It has a wingspan of 3 inches and is our only underwing with both black and white bands on the underwings. The forewing varies in color from black or gray to white. During the day this large moth rests well camouflaged on tree bark with its colorful hindwings folded beneath cryptic forewings. There is one generation a year. The White Underwing ranges across the northern United States and southern Canada.

Smooth With Knobs or Bumps

DARLING UNDERWING   To 4 inches

This underwing larva is a mottled gray-brown color. It has *yellow bumps or knobs* on each segment, including 1 *larger* knob on the eighth segment. Like the White Underwing larva, it has white *fringe* on its undersides. The head has whitish markings on top and in front. The larva eats willows, preferring black willow. It overwinters as an egg.

The adult moth's forewings are brown; the hindwings are bright pink with black bands. Its wingspan is about 2½ inches. The Darling Underwing is found in northern New England and Canada, south to Florida and west to South Dakota and Texas.

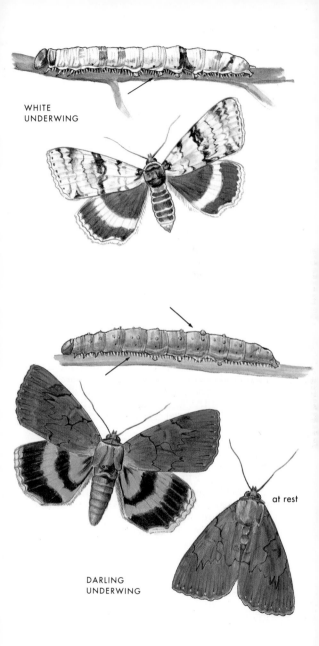

WHITE
UNDERWING

DARLING
UNDERWING

at rest

47

Smooth With Knobs or Bumps

VICEROY
RED-SPOTTED PURPLE
WHITE ADMIRAL   To 1¼ inches

As the larvae of these three closely related
butterflies are identical, they are presented
here as one. These caterpillars look *like bird
droppings*, which may protect them from
predators. They are brown or greenish with
a *whitish saddle*. They have *horns with
short, thick branches* and *bumps on 4 segments*. The Viceroy larva eats willow, poplar,
and fruit trees; the Red-Spotted Purple eats
willow, poplar, and wild cherry; and the
White Admiral eats birch, willow, and
poplar. Either the larva or chrysalis overwinters. The chrysalis also resembles a bird
dropping.

Although the larvae look alike, the three
adult butterflies are very different. It has
long been thought that the orange and black
coloration of the Viceroy protects it from
birds because it resembles the toxic Monarch butterfly, which receives toxins from
its food plant. Some scientists now believe
that Viceroys in certain areas might manu-
facture their own toxic chemical. The black
line across the Viceroy's hindwing is not
present on the larger Monarch. The Viceroy
flies in open areas, meadows, and wetland
edges.

The Red-spotted Purple and the White
Admiral are actually two forms of the same
species. The White Admiral is a northern
butterfly that likes high hardwood forests
and mountainsides; the Red-Spotted Purple
is southern, preferring low, open areas and
a warm climate. The Red-spotted Purple
resembles the toxic Pipe-vine Swallowtail;
birds avoid both, but only the Pipe-vine is
poisonous.

All three butterflies are found throughout
the eastern United States, west to Texas and
parts of the Southwest, north through the
central United States and into Canada.
Each has a wingspan of about 3 inches.

VICEROY

RED-SPOTTED
PURPLE

WHITE
ADMIRAL

49

Smooth With Knobs or Bumps

The larvae described on pages 50–59 are members of the giant silkworm family. The caterpillars are generally large and easy to rear. The moths are large and attractively colored and are often mistaken for butterflies. The Silkworm Moth, which was once raised for its silk threads, is in a different family.

CECROPIA MOTH   To 4 inches

This caterpillar changes color as it grows. The young larva is black and covered with bristles, later turning orange, then greenish. The mature larva is greenish and covered with *blue and yellow knobs* called tubercles. There are also *4 red tubercles* on 2 segments near the head. The Cecropia larva prefers silver maple, but also feeds heavily on wild cherry, maple, apple, poplar, oak, sassafras, gray birch, dogwood, and many other deciduous trees. When ready to pupate, the mature caterpillar builds a silk cocoon around itself before it becomes a pupa. The pupa overwinters inside its cocoon. The cocoon is built lengthwise along a twig of its food plant. It may look like the one in the illustration, or it may be smaller and more tightly constructed. The female moth lays eggs in groups, as shown, on the undersides of leaves.

This colorful moth has beautiful markings, making it popular with collectors. The male, larger than the female, is the largest moth in North America, with a wingspan of 4–6 inches. The male's antennae are larger and more feathery than the female's. The male's large antennae are sense organs used to receive the chemicals given off by the female. In this way, males can locate females in the dark of night. Also called the Robin Moth, the Cecropia is found east of the Rocky Mountains and in southern Canada.

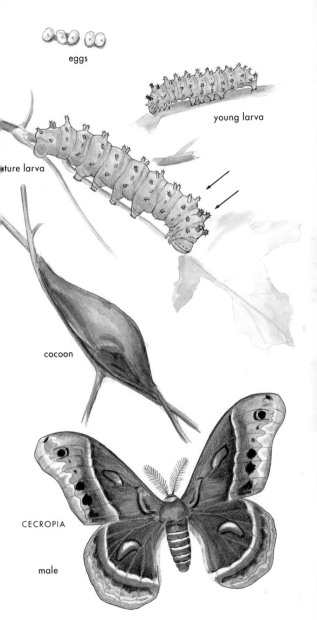

eggs

young larva

ture larva

cocoon

CECROPIA

male

51

Smooth With Knobs or Bumps

LUNA MOTH   **To 2³/₄ inches**

The larva of the Luna Moth is smooth and
fleshy, a translucent light green with a *light
yellow line* along each side. It has many *red
or orange short round knobs* called tubercles. It has some fine hairs. The head may
be brown or green. The caterpillar resembles
that of the Polyphemus Moth (page 56), but
adults of the two species are very different.
The Luna Moth larva prefers hickory but will
also feed on other nut trees, willow, maple,
persimmon, sweetgum, birch, oak, alder,
and beech. The cocoon is made from leaves
and covers the pupa loosely; the overwintering pupa can sometimes be heard moving
inside the cocoon. It is usually hidden under
the loose leaves beneath the food plant but
is not easy to find. The large eggs are oval
and dark, laid in irregular groups. There are
two generations a year, except in the North,
where the growing season is short.

The night-flying Luna Moth, or Moon
Moth, is very popular with collectors for its
spectacular colors and remarkably long tails
(it is unrelated to the swallowtail butterflies). Its wingspan can reach 5 inches. The
color of the wing edges varies according to
the season: Adults that emerge in spring
have pink-purple margins, while those
emerging in summer have yellow wing margins. All the colors will fade when exposed to
light. The illustrations show the moth at
rest, with wings folded, and with the wings
spread, exposing the transparent spots on
each wing. The Luna Moth is found in deciduous forests of the eastern United States;
look for it at night in late spring and early
summer.

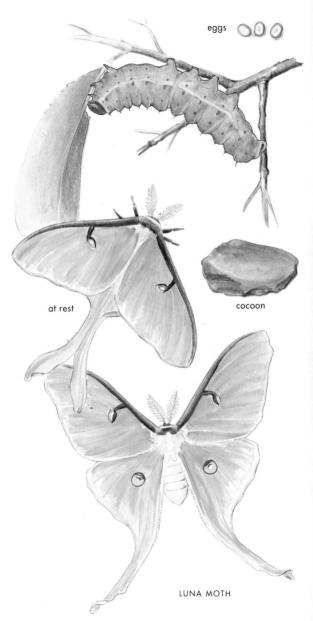

eggs

at rest

cocoon

LUNA MOTH

53

Smooth With Knobs or Bumps

PROMETHEA MOTH   To 3 inches

The caterpillar is blue-green and mostly smooth, with *rows of different colored knobs*, or tubercles. The *2 long pairs of tubercles* on the front segments are *orange or red*. The tubercle at the rear segment is *yellow*. Each of these long tubercles has a black ring at its base. Other short tubercles along the body are *black*. The larva prefers to eat wild cherry and spicebush but also feeds on a variety of deciduous trees and shrubs including maple, apple, ash, basswood, tulip tree, sweetgum, birch, and sassafras. Only young larvae feed close together. The pupa overwinters inside a silk cocoon that usually hangs down from the host plant within a curled leaf. It remains attached to the twig after the leaves have fallen and is easily spotted in winter and spring by collectors and predators. It was once considered possible to raise this moth commercially for the caterpillars' silk. Eggs are white, flat ovals laid in a row.

The male and female of this species, also called the Spicebush Silk Moth, look quite different. The female has a wingspan of up to 6 inches and is red-brown with colorful patterns. The male is smaller and dark colored and flies during the late afternoon to mate. Many males may swarm around a female. The Promethea Moth is found in the eastern half of the United States and into Canada. Southern populations have two generations per season.

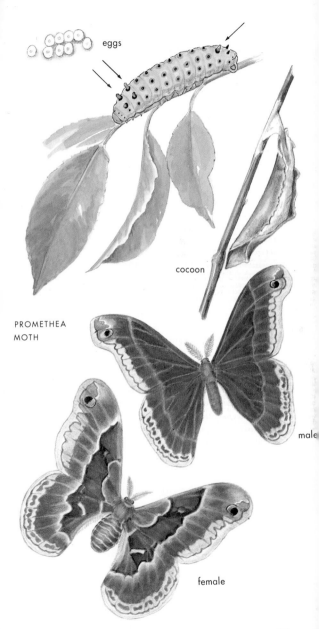

eggs

cocoon

PROMETHEA
MOTH

male

female

55

Smooth With Knobs or Bumps

POLYPHEMUS MOTH   **To 3 inches**

This may be the most common of the giant silkmoths. The eggs are large, rounded, and flattened at the top, with rust brown bands around the outsides. The larva is smooth and green with numerous *red, orange, or yellow knobs*, or tubercles, from which extend *short hairs*. The spiracles are *red*. Behind each one is a vertical yellow stripe. The larva prefers to eat oak, but also feeds on hickory, elm, maple, birch, linden, willow, and chestnut. Here it is shown devouring a maple leaf. Its large size and slow movements make it an easy meal for predators. When ready to pupate, the mature larva wraps a leaf of the host plant around itself and binds it with silk. The tough, rounded cocoon may be found attached to the host plant or in the leaf litter at the base of the tree. This is an ideal cater-pillar to rear indoors, especially for children, because it is large, harmless, and easily handled, it will eat a wide variety of food plants, and it transforms into a beautiful adult moth.

The Polyphemus Moth has a wingspan of up to 6 inches. It is named after a one-eyed giant in Greek mythology for the large eye-spots on its hindwings, which may serve to frighten predators. There are also small transparent spots on the forewings. The colors of the adults can vary slightly. The moth illustrated with wings spread is the usual orange-tan color. Above it, with wings slightly folded, is a green-tan variation. This variance could be due to temperature, food quality, or geographic location. The adults are often attracted to lights in midsummer. The male is larger than the female, and his antennae are more feathery. The Polyphe-mus is found throughout the continental United States.

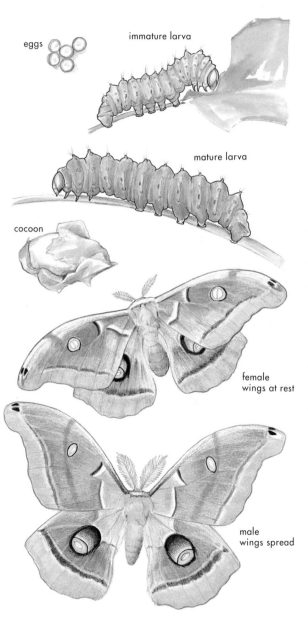

eggs

immature larva

mature larva

cocoon

female
wings at rest

male
wings spread

POLYPHEMUS MOTH

Smooth With Knobs or Bumps

HICKORY HORNED DEVIL   To 6 inches

This larva can be intimidating because of its very large size, its spines, and its spiky horns. But though its appearance is less than friendly, it is a harmless caterpillar, and its spines are not irritating to skin. It usually has a green body with *spines*. Segments behind the head have long, curved, *black and orange horns* with short hairs. It eats hickory, pecan, walnut, sweetgum, persimmon, and sumac. The larva matures in late summer and may be seen away from its food plant before pupation. The hardened, dark pupa overwinters in the soil, sometimes through two winters.

The adult is called the **Royal Walnut Moth** or the Regal Moth. It has orange stripes and yellow markings. Adults may be attracted to lights at night. The female, with a wingspan of up to $5\frac{1}{2}$ inches, is larger than the male. It is found in the eastern United States, more commonly in the South.

IMPERIAL MOTH   To 4 inches

This silkworm larva varies in color; although it is usually green, it can be brown or tan. It has several *orange or yellow knobs,* or tubercles, and some *long, fine hairs.* The yellow or white spiracles are *large* and easy to see. It eats the foliage of many trees, including oak, maple, linden, birch, elm, walnut, cedar, and pine.The large, hard pupa has a tail that is flattened and forked. It overwinters in soil or leaf litter.

Male and female Imperial Moths look different; the female is larger, with a wingspan up to $5\frac{1}{2}$ inches, and is more yellow overall than the male. The male bears stronger markings, particularly more pink along the outer edge of his forewings. The Imperial Moth is found east of the Rocky Mountains and in southeast Canada. One generation occurs in a growing season.

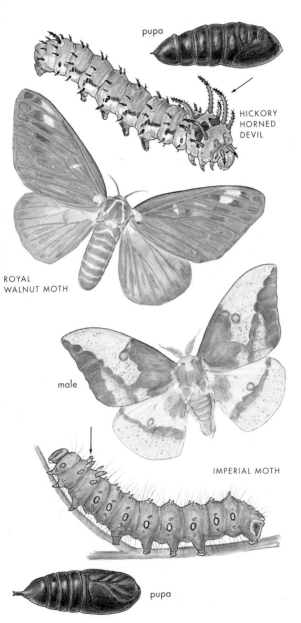

pupa

HICKORY
HORNED
DEVIL

ROYAL
WALNUT MOTH

male

IMPERIAL MOTH

pupa

59

Smooth With Rear Horn or "Tail"

SPHINX MOTHS or HAWK MOTHS

The larvae of the sphinx moths are usually large, smooth, and green, with diagonal stripes on their sides. Most caterpillars in this family have a distinct *horn or "tail"* on the top of the rear segment. This harmless horn may make the caterpillar look frightening to invaders, especially when it rears up its front end in the "sphinxlike" pose that gives the family its common name. The wings of the adult moths are narrow, and their bodies are robust but tapered. They are among the fastest fliers of the butterflies and moths. Their proboscis, or hollow tongue, is quite long, and sphinx moths hovering in flight to feed from flowers can be mistaken for hummingbirds or bumblebees. The pupa is large and may have a tongue case that looks like a jug handle near the head. The pupa overwinters in the soil. Many sphinx moths are easily reared.

WAVED SPHINX   To 4 inches

The caterpillar is shown in two color forms; the green form is more common. The reddish form could be due to genetic variation or perhaps to geographic isolation. The green form turns distinctly pink before pupating in the soil. Both forms have a *pink stripe* on the sides of the head and *diagonal stripes* on the sides of the body. Notice the reddish *fleshy horn* on the rear segment. The larva feeds on ash, privet, oak, lilac, and hawthorn. Here it is shown on privet.

The Waved Sphinx's name refers to the wavy markings on the forewing. The adult is large, up to 4 inches across its spread wings, and is lighter in color than many other hawk moths. It is seen throughout eastern North America and may be seen flying around outdoor lights in the summer.

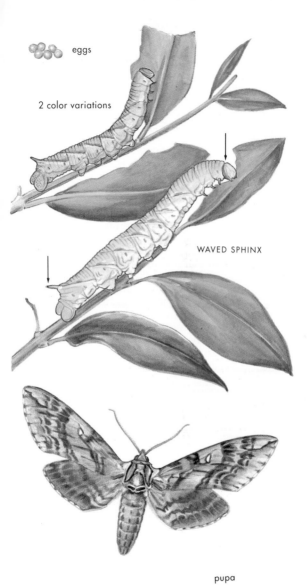

eggs

2 color variations

WAVED SPHINX

pupa

61

Smooth With Rear Horn or "Tail"

TOBACCO HORNWORM   To 6 inches

The Tobacco Hornworm is an economically significant pest because of its voracious appetite for tobacco, tomato, potato, and other related crops. An abundant food supply will nourish the larva to maturity at a startling 6 inches in length. It is green or sometimes green-black with *7 whitish diagonal lines* above the spiracles along each side. The rear horn is *red*. It is the larva of the **Carolina Sphinx Moth.**

The **Tomato Hornworm**, larva of the **Five-spotted Hawk Moth,** is very similar, but its rear horn is green and black, and its diagonal stripes form an angle around the spiracles instead of the straight lines of the Tobacco Hornworm. It feeds on the same plants. Both larvae are often victims of parasitic wasps; the wasps' white cocoons may protrude from the host caterpillar's body.

With wings 4 1/2 inches across, the adult moths look alike, with yellow-orange spots along the sides of their abdomens. They are common in most of the United States. The overwintering pupa has a tongue case, or "jug handle," near the head.

LAUREL SPHINX   To 4 inches

The larva feeds on laurel as well as privet, poplar, lilac, ashes, and fringe trees. Here it is shown on privet. Its most distinguishing feature is the *blue rear horn* with *black markings.* The horn is present in all instars, even the youngest larvae, and can be found on shed skins. The diagonal lines above the spiracles are *blue-black over yellow.* On the side of its head is a dark band.

The adult moth has a wingspan of up to 4 1/2 inches. The pupa, which overwinters in loose soil, has a short tongue case close to the body. The egg is large, round, and light green. The Laurel Sphinx ranges from southeast Canada to Florida and west to Arkansas and Louisiana. It is more common in the northern reaches of its range.

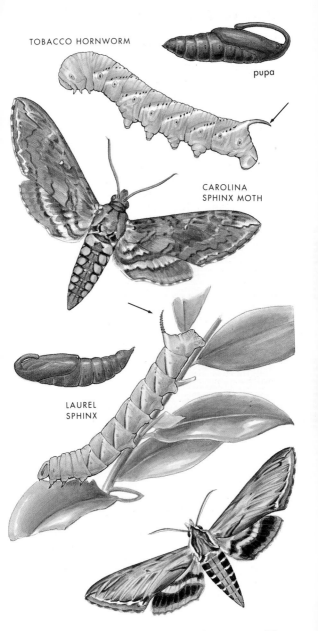

TOBACCO HORNWORM

pupa

CAROLINA
SPHINX MOTH

LAUREL
SPHINX

63

Smooth With Rear Horn or "Tail"

HOG SPHINX   To 3 inches

This colorful sphinx moth caterpillar can
retract the first two segments of its body
into the third segment, which is *enlarged* to
accommodate this remarkable behavior. The
swollen segment gives the larva its "hoglike"
jowls. The last segment has a horn or "tail."
The larva feeds upon grape and Virginia
creeper. It is a colorful caterpillar with *blues*
and *greens* that fade into muddy colors just
before the insect pupates. It pupates in light
soil, surrounding itself with a thin net of
loose silk that serves to keep the soil away.

The adult hawk moth, also called Virginia
Creeper Sphinx Moth, has orange-brown
wings about 2½ inches across. It is shown
here with wings extended and at rest. The
Hog Sphinx is an eastern species, but its
range extends west to Kansas and Iowa.

HUMMINGBIRD CLEARWING   To 2 inches

The larva feeds upon viburnum, hawthorn,
honeysuckle, and some fruit trees. It varies
in color but is generally yellow-green with
white stripes along its back. There are *pink*
markings on the underside of the last two
segments. The last segment has a *horn* or
"tail." This caterpillar is shorter and stockier
than most other sphinx moth larvae.

The adult Hummingbird Clearwing is a
day flier and is attracted to many fragrant
garden flowers. Because its wings move so
quickly that it can hover as it sips nectar
with its long tongue, or proboscis, many
people mistake this moth for a hummingbird or a bumblebee. It has a wingspan of 2
inches or more. It is common throughout
the East into Canada. There is one generation per year in the North, two in the South.

The scales wear off large areas of the forewings of clearwing moths, resulting in the
transparent "windows."

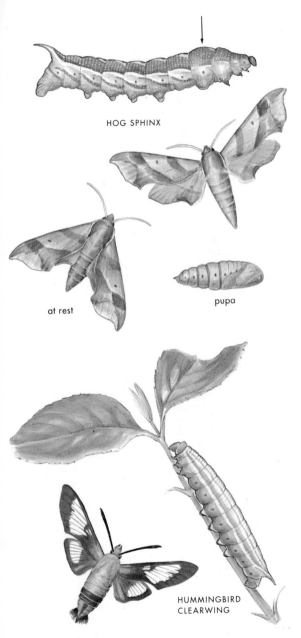

HOG SPHINX

at rest

pupa

HUMMINGBIRD
CLEARWING

65

Smooth With Rear Horn or "Tail"

CATALPA WORM   To 2³/₄ inches

Large numbers of Catalpa Worms can thor-
oughly defoliate catalpa trees, particularly
in southern areas. In the Southeast, the caterpillar is used as fish bait. The larva has
two color forms, dark and light, both with a
rear horn. The dark form is shown in the
illustration. The lighter form is pinkish
below and on the sides; the back is *dark*
(but lighter than in the dark form) and
somewhat *striped*. Catalpa Worms are often
parasitized by wasps. The wasps inject
numerous eggs into the caterpillar body.
The wasp larvae feed inside it until they are
mature and ready to pupate. Their small
silken cocoons protrude from the caterpillar's body, leaving it a weak shell. Wasp
adults emerge, and the caterpillar dies.

The adult **Catalpa Sphinx** has a wingspan
of about 3 inches. It lives in the eastern
United States north to southern Michigan
and New York, west to the Mississippi
Valley.

WHITE-LINED SPHINX   To 3¹/₄ inches

The larva has many color varieties, including a green form and a black form. The most
common one is shown, a black caterpillar
with *yellow spots,* a pink or yellow line along
the back, and *yellow lines* on the sides. The
areas around the spiracles (circular spots
on the sides through which the caterpillar
"breathes") are black or yellow, as is the
fleshy horn. It eats Virginia creeper and
grape, purslane, apple, and many other
plants.

The day-flying adult is also called the
Striped Hawk Moth for the bold white or
yellow diagonal bands on its forewings. The
egg, like most in the sphinx moth family, is
oval and pale green. This species overwinters as a pupa in the soil. Its range extends
throughout the United States and south
into Central America.

CATALPA
WORM

CATALPA
SPHINX

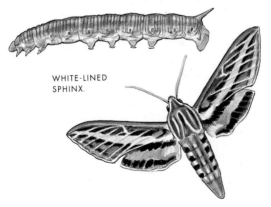

WHITE-LINED
SPHINX.

67

Smooth With Rear Horn or "Tail"

PANDORUS SPHINX   To 3$\frac{1}{4}$ inches

For the first few instars, the young caterpillar is green in color and has a long "horn" or tail on its last segment. The horn may be curled. After the third molting, the horn disappears, leaving a black, shiny, *eyelike bump.* The larva may stay green or turn brown. The *spots* on the sides of the Pandorus larva are *oval,* whereas on the similar Achemon Sphinx larva, the spots are angular. The Pandorus larva can retract its first two body segments into its third when disturbed or at rest. It usually walks and feeds with the segments extended, feeding on grapes and Virginia creeper and ampelopsis, or blueberry climber. It is common throughout the East, west to Kansas and Texas.

The adult Pandorus Sphinx has wings more than 4 inches across. It is olive green in color with pink markings on its wings.

ABBOT'S SPHINX   To 2$\frac{1}{2}$ inches

The young larva is white; the mature larva has two common color forms, as shown in the illustrations: tan or brown with *greenish saddles* on each segment or cream or brownish with darker stripes along the back and sides. It *lacks* a horn but has an *eyelike bump* on its last segment. When disturbed, it thrashes its head and forebody from side to side as a defensive ruse. The Abbot's Sphinx caterpillar feeds on grape, ampelopsis, and Virginia creeper. It feeds at night. During the day, when it is still, the larva's coloration allows it to blend in with the bark of the vines. Look for the caterpillar on vines near the ground. The bare pupa, which overwinters, is shiny and red-brown in color.

The adult moth has a wingspan of just under three inches. Its dark wings have scalloped margins, and the hindwings have yellow bands. The dark body resembles that of a bumblebee. It occurs in the East, west to Kansas and Texas.

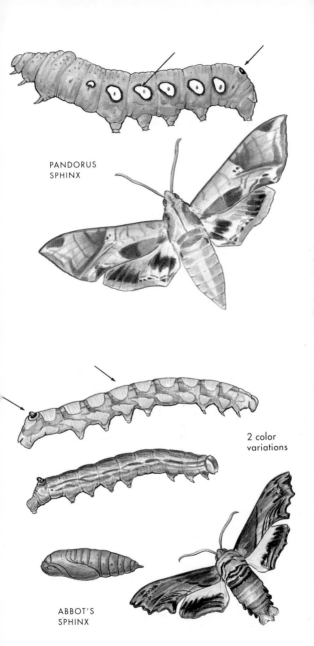

PANDORUS
SPHINX

2 color
variations

ABBOT'S
SPHINX

Smooth With Fleshy Filaments

MONARCH   **To 2³/₄ inches**

This larva can be found on the undersides of milkweed leaves, on which it feeds. The caterpillar absorbs toxins from this plant and is distasteful to birds. It is *white* with *yellow and black stripes*. There is one pair of long, fleshy, *black filaments* at the head and another pair at the tail. The roundish chrysalis is green or blue with gold markings. The egg is round and pale green or cream, laid singly under a leaf of the host plant.

The large Monarch, with its orange and black stripes, is probably our most familiar butterfly. It sips nectar from wildflowers and garden flowers. The Monarch is famous for its remarkable migrations. During the summer, the Monarch lives and reproduces throughout the United States. Fall-emerging adults migrate south to California and Mexico, where the butterflies can be seen roosting in trees in great numbers. In the spring, those same butterflies mate and begin the return trip north, the females laying eggs on young milkweed plants as they go. A migrating adult survives both the trip south and the return trip north in its 9-month lifetime. Since Monarchs pass the winter as adults, not eggs or pupae, northern areas must be recolonized every year.

QUEEN   **To 2¹/₂ inches**

The Queen's larva resembles the Monarch's but is *darker brown,* and the Queen's *yellow* markings are *not* continuous bands. This caterpillar has *3 pairs* of black filaments, one pair each on the front, middle, and rear of the body. Like the Monarch, the Queen is a milkweed butterfly. Both the caterpillar, which feeds on milkweeds, and the adult are toxic to predators such as birds.

The adult Queen's wings are yellow-brown with a dark brown border and white markings. The chrysalis is similar to the Monarch's. The Queen ranges through the southern areas of the United States, from Florida to southern California.

70

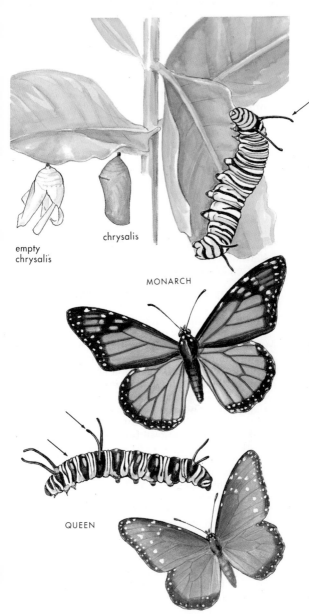

empty
chrysalis

chrysalis

MONARCH

QUEEN

71

Smooth With Fleshy Filaments

GREEN-STRIPED MAPLEWORM

To 1^1/$_3$ inches

The caterpillar is named for its markings and food plant. It prefers Silver Maple but will eat any of the maples and also oaks. The larva is covered with short spiked protuberances and has one pair of black long *fleshy filaments* near the head. It is light green with several *darker green stripes* lengthwise along the body and *pink* markings on the underside near the rear. The head is brownish red. Large groups of these larvae feeding together can strip maple trees. The mapleworm overwinters as a pupa in light soil. The small eggs are bright yellow.

The adult, the **Rosy Maple Moth,** is a small, pretty moth colored in pale yellow and pink. It may be seen fluttering around lights on summer evenings in the eastern United States.

PINK-STRIPED OAKWORM

To 2 inches

The Pink-striped Oakworm is named for its markings and its host; it prefers red oaks and is considered a pest to the trees. The caterpillar may be specked with white dots and is green or tan with *pink stripes*. It has a pair of black *fleshy filaments* extending from the second segment behind the head. It has many short, spiked protuberances on its body. It overwinters in the soil as a pupa. The illustration shows the pupa after the moth has emerged.

The adult female Pink-striped Oakworm Moth, with a wingspan of over 2 inches, is larger than the male, and her wings are more rounded than his narrow, angular ones, which may be marked with black speckles. Both male and female have the prominent white spot near the center of the forewing. The male may have transparent areas just outside the spots. This is also an eastern species.

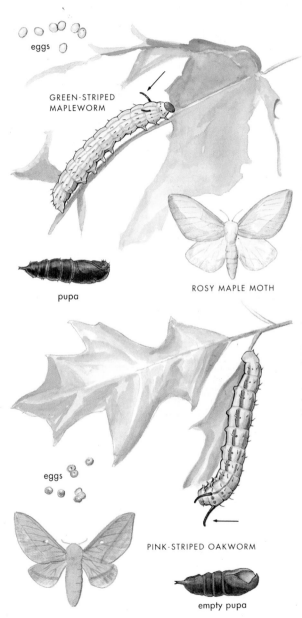

eggs

GREEN-STRIPED MAPLEWORM

pupa

ROSY MAPLE MOTH

eggs

PINK-STRIPED OAKWORM

empty pupa

73

Smooth With Fleshy Filaments

BLACK-ETCHED PROMINENT   To 1¹/₄ inches

The young larva is black, but later instars (the stages between molts) are colored like the one in the illustration. Notice the *bump* on the third segment behind the head and brown saddle on the middle segments. This caterpillar uses posture, coloration, and even chemicals for defense. When disturbed, the larva rears up both ends and retracts its head to show an alarming *red and black "face."* From the pair of fleshy filaments on the last body segment, it can *extrude pink tentacles* that give out a foul odor. It feeds upon poplar, willow, and wild cherry, usually near rivers and lakes. The pupating larva builds a tough brown cocoon using its silk and parts of the food plant to protect the overwintering pupa.

The small adult moth is found throughout the United States and into Mexico.

PIPE-VINE SWALLOWTAIL   To 2 inches

The mature caterpillar can be startling when seen for the first time. It is *purplish black,* and all of its body segments are studded with *blackish red, fleshy filaments.* As the caterpillar walks, it explores its path with the *long filaments* on the first segments and some on the sides. Predators avoid the Pipe-vine Swallowtail caterpillar, as well as the adult butterfly; the larva absorbs toxins from its food plants, which include pipe-vines, Dutchman's-pipe, and Virginia Snakeroot. This species is so distasteful that predators avoid any butterfly that looks like the Pipe-vine Swallowtail—even edible ones.

Eggs are rust colored. The illustration shows how the chrysalis is supported across its girth by a silken thread attached to a stem for overwintering. Notice the yellow rounded areas that protrude from the chrysalis. This butterfly, with a wingspan of 4¹/₂ inches, is found throughout the East and to California but is absent from the Northwest.

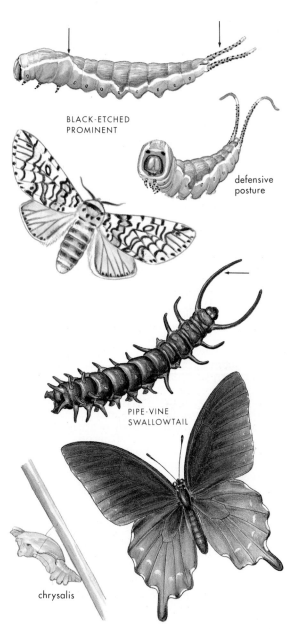

BLACK-ETCHED
PROMINENT

defensive
posture

PIPE-VINE
SWALLOWTAIL

chrysalis

Sluglike

GRAY HAIRSTREAK   To ⅝ inch

The larva is called the Bean Lycaenid or
Cotton Square Borer, as it feeds upon beans
and other legumes including clover and
alfalfa. It eats the flowers and bores into the
fruits and seeds, and it is often found in veg-
etable gardens. The larva is small—notice its
size compared to the clover blossom in the
illustration. Its head and legs are hard to
see beneath its stout, thick body which
varies in color from green to cream to pink
(2 color forms are illustrated). It is covered
with short brown hairs. The pupa shape is
also characteristic of the hairstreak butterfly family.

Also called the Common Hairstreak, the
adult is small, about 1 inch across, and a
quick, low flier. The female is browner than
the male; both have a pair of tiny tails on
each hindwing, one longer than the other.
When at rest, it folds its wings upright,
showing the light gray underwings. The
male is territorial and protects his area from
other males. The butterfly is commonly
found in fields, meadows, and wastelands
throughout North America except extreme
northern Canada.

GREAT PURPLE HAIRSTREAK   To ½ inch

The larva eats the flowers and leaves of mis-
tletoe. Its short, thick body is green with a
tinge of red, has a yellow stripe on each side,
and is covered with short, fine, orange hairs.
There are two generations a year. The pupa
overwinters.

The adult butterfly is about 1½ inches
across. It is colored with iridescent blues
and greens that change depending on the
angle of light that strikes them. The male is
generally more brightly colored than the
female. Their underwings are brown. There
are usually two tails on each hindwing. This
hairstreak is found in the southern half of
the United States, from New Jersey west to
California and south into Mexico.

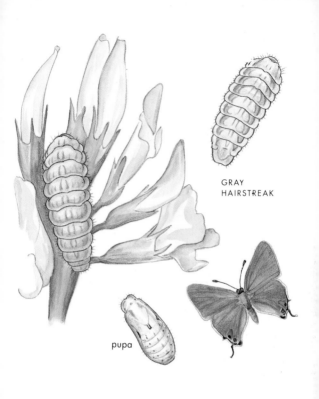

GRAY
HAIRSTREAK

pupa

GREAT PURPLE HAIRSTREAK

female

male

77

Sluglike

Coppers are small butterflies found primarily in the Northern Hemisphere. Most of the adults have an iridescent, coppery sheen. The larvae are short and thick.

PURPLISH COPPER   To 1/2 inch

The sluglike Purplish Copper larva is green and covered with short white hairs. Side and back stripes are usually yellowish, sometimes reddish. It feeds on dock, knotweed, and baby's-breath found in wet fields and damp meadows and along streams and roads. Its range is mostly western, north to Alaska and northeast to the Great Lake states. The half-grown larva overwinters. The female adult is larger than the male and shows less purple.

SMALL COPPER   To 1/2 inch

The larva of the Small Copper can be found in meadows and weedy areas where its food plant, sheep sorrel, grows. The caterpillar may be light green, yellow-green, or red. The pupa overwinters. The adult, also called the American Copper, is found in the eastern United States and Canada.

HARVESTER   To 1/2 inch

This caterpillar is remarkable because its eating habits are actually beneficial to its host plants: It is the only carnivorous butterfly larva in North America. It eats woolly aphids, which suck the juices out of alders, beech, ash, viburnum, hawthorn, and other trees and shrubs. To protect itself from ants that eat the "honeydew" secreted by aphids, the caterpillar covers itself with aphid skins and silk. Without its covering it is gray with many hairs. The pupa, which resembles a monkey's face, overwinters. Whitish eggs are laid among the woolly aphids.

The adult Harvester sips on aphid honeydew, as well as sap, mud, and dung. Its wings are dark brown with orange markings. It is found in the eastern United States, west to central Texas.

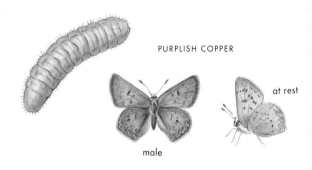

PURPLISH COPPER

male   at rest

SMALL COPPER

at rest

HARVESTER

at rest

79

Sluglike

Blues, like their relatives the hairstreaks and coppers, are small butterflies with lustrous wings. Their caterpillars are small and slug-like in shape, and many are attended by ants. These caterpillars secrete a sweet liquid called honeydew that the ants like. The caterpillars benefit because the ants defend the caterpillars against predators.

SPRING AZURE   To $1/2$ inch

The larva is small; it can be green, whitish, or rose-colored, and has a small head. Its host plants include flowers and flower buds of dogwood, snakeroot, viburnum, blueberry, and spirea. Ants tend this larva for its honeydew. The overwintering pupa is hidden on stems or in leaf litter.

The Spring Azure is seen throughout most of North America. The appearance of the 1-inch adults in open woods, fields, and roadsides is one of the first signs of spring.

EASTERN TAILED BLUE   To $1/2$ inch
WESTERN TAILED BLUE

The larvae of these two closely related species are similar: small, downy, and dark green, with a brownish, reddish, or greenish *stripe along the back* and a white and red stripe along each side. The Eastern Tailed Blue is shown. They eat legumes including clovers, trefoils, and peas.

These small butterflies are among the most common in the United States. The Eastern is found generally east of the Rockies; the Western is found west of the Rockies with some overlapping in the central states.

MELISSA BLUE   To $1/2$ inch

As do most blues larvae, this one eats flowers or leaves of legumes. The Melissa Blue larva is green, *darker on the back* and yellowish on the sides.

Melissa Blue is mainly a western species. A northeastern subspecies called the Karner Blue is an endangered species that lives in pine bush habitats.

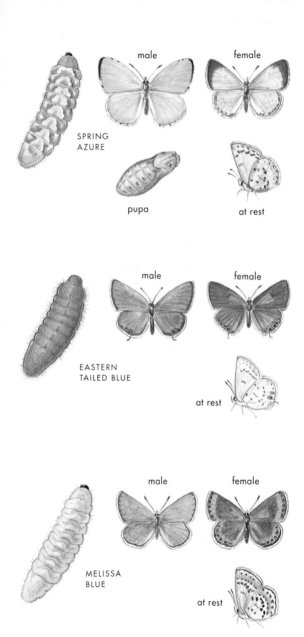

male

female

SPRING
AZURE

pupa

at rest

male

female

EASTERN
TAILED BLUE

at rest

male

female

MELISSA
BLUE

at rest

Hairy

GYPSY MOTH   To 2 inches

The Gypsy Moth caterpillar is a notorious pest because it defoliates deciduous trees and shrubs. Large populations can strip huge tracts of forest—the munching sound can be heard by a careful listener. The Gypsy Moth is not native to the United States; it was brought to Massachusetts from Europe around 1868 for the purpose of silk production, though it produces a trivial amount of silk. Some of the moths escaped, and the species became well established in the wild.

The larva favors trees and shrubs such as apples, cherries, oaks, and willows. If these food plants are not available, however, the caterpillar will eat many other less favored trees and shrubs. It can travel to fresh food plants by lowering itself from its long silk thread, then allowing the wind to blow it to adjacent trees. Young larvae are black. The mature caterpillar is gray with *white side stripes,* and there is an *orange stripe* along its back. Its body is covered with *long hairs.* Most noticeable are the *round, colored bumps,* or tubercles, on each segment. Those on the first 5 segments behind the head are *blue,* the remaining seven are *red.* The side tubercles are yellow. The head is large, with black and yellow markings. Hairs from the larva can be seen on the dark brown pupa, which is usually found on the trunk of a host plant.

The female adult, with a wingspan of $2\frac{1}{2}$ inches, is whitish with dark markings. The male is smaller and brown-colored. The female lays eggs on tree trunks in creamy brown masses that overwinter and hatch the following spring. Populations fluctuate, but the Gypsy Moth has extended its range from New England north to Canada, south to Florida, and west to Missouri and Michigan. Scientists and foresters are looking for ways to control this pest with natural predators.

GYPSY MOTH

male

female

egg mass

pupa

83

Hairy

EASTERN TENT CATERPILLAR

To 2 1/4 inches

Colonies of the Eastern Tent Caterpillar are easily recognized by the protective tents of silk that the caterpillars build in their host plants. The larvae feed on young green leaves in the outer branches of wild cherry, apple, pear, and other fruit trees. Watch for the tented crotches in wild cherry as spring progresses. Many caterpillars take shelter in each nest. They feed outside of their tents, stripping the leaves from the branches extending from the nest. By the end of spring that portion of the tree is bare, and by summer, wind and rain have broken down the tent. This is a colorful caterpillar: Black stripes border one white stripe along its back, and the sides are *blue* with an *orange stripe.* Each segment has white and black *eyespots*, and the caterpillar is covered with *orange-yellow hairs.* The pupating caterpillar leaves the tent and makes a thin, creamy to pale yellow cocoon in a crack or crevice near or away from the host plant. The pupa might be visible inside the spun cocoon.

The adult moth is light brown to brownish red with lighter diagonal stripes on the forewings. It has a wingspan of 1 3/4 inches. It is shown with wings folded back against the body, its natural resting posture. Eggs are laid in a mass that hardens around a twig for overwintering. The egg mass is shown unhatched, as it would appear in winter, and hatched, as in spring. It is common throughout the eastern and central states to the Rocky Mountains. The similar Western Tent Caterpillar is found along the West Coast.

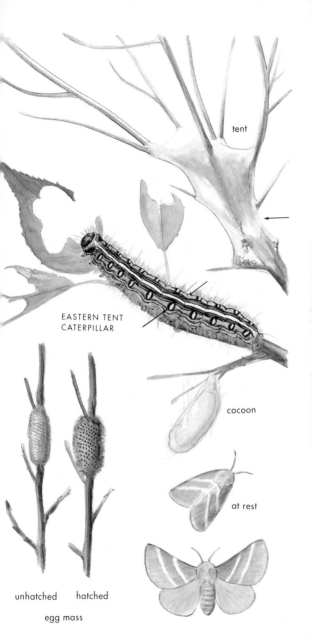

tent

EASTERN TENT
CATERPILLAR

cocoon

at rest

unhatched   hatched

egg mass

85

Hairy

DELICATE CYCNIA   To 1 inch

This caterpillar feeds almost exclusively on
dogbane, a milkweed relative, but it will eat
other milkweeds and also Indian hemp. The
larva is covered with *soft white or gray hairs*.
When it is ready to pupate, the larva makes
its cocoon out of the hairs. The dark pupa
can be seen inside the thin, transparent
silken cocoon.

Also called the Dogbane Tiger Moth, the
adult is a medium-sized, creamy or white
moth with varying amounts of yellow bor-
dering the edge of the forewing. The illustration shows the adult with wings spread and
also with wings folded in its natural resting
posture. It is found throughout the United
States.

YELLOW BEAR   To 2 inches

The commonly seen caterpillar feeds on
plants such as birch, maple, cabbage, corn,
and many other trees and garden crops. It
can be *very hairy*, but if you look closely you
might still see the separate segments of its
body. The color of the hairs may be white,
yellow, orange, or reddish. The pupa is protected in a soft cocoon made of hairs from
the caterpillar, perhaps including bits of
plant material.

The adult **Virginian Tiger Moth,** with a
wingspan of about 2 inches, is less commonly seen than its caterpillar. Its wings are
white with small black markings, and there
are black spots and yellow side stripes on
the abdomen. The forelegs are white, yellow,
and black. This tiger moth is common
throughout the United States.

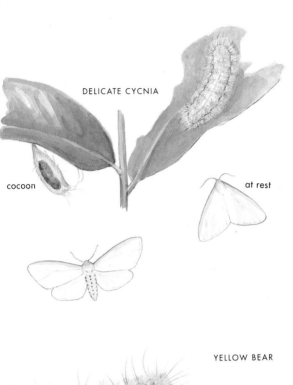

DELICATE CYCNIA

cocoon

at rest

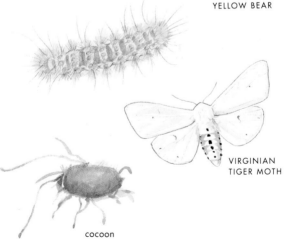

YELLOW BEAR

VIRGINIAN
TIGER MOTH

cocoon

Hairy

VIRGINIA CTENUCHA   To 1 1/4 inches

This caterpillar's black or cream-colored hairs are in short *clumps*. The body is generally brown, lighter on its back, and the head is reddish brown and black. It feeds on grasses, irises and sedges.

The adult moth has a wingspan of about 2 inches. Its wings are brown with white along the edges, but its body is metallic blue with an orange head and collar. This is our country's largest wasp moth, so named because it looks somewhat like a wasp when it feeds at flowers by day. It is common in most of the eastern United States.

FALL WEBWORM   To 1 1/2 inches

You can easily spot this pest because it makes messy webs in the leaves and branches of its host plant in late summer and fall. Like the Tent Caterpillar, it feeds in colonies, using the webbing as a shelter. Heavy feeding can seriously damage many species of trees, including willow, ash, hickory, maple, oak, walnut, and apple. The Fall Webworm is covered with long, fine hairs that may be white or yellow. There are *black spots* on each body segment.

The adult Fall Webworm Moth is small (1 1/2 inches) and white with brown or black spots on the forewing. It is common throughout most of the United States.

NORTHERN METALMARK   To 3/4 inch

The greenish larva's most distinguishing feature is the *unique growth habit* of its long white hairs: straight up in one row along the back and outward in one row along each side. Shown are top and side views. The caterpillar feeds on ragworts. Winter is passed by the caterpillar. The pupa is covered with long, whitish hairs from the larva.

The small adult butterfly often perches with its wings open. It is found in dry, hilly areas or open woodlands in the Northeast, but it is usually isolated and local.

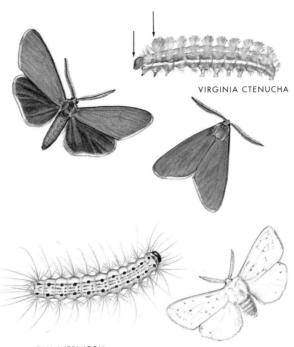

VIRGINIA CTENUCHA

FALL WEBWORM

NORTHERN METALMARK

side view

89

Hairy

The two caterpillars shown here belong to a group with an unusual defensive posture. If one of these larvae is threatened, it arches its head and rear segments backward. Although they are grouped with other hairy caterpillars, these are only sparsely covered with hairs.

AZALEA CATERPILLAR To 2¹/₄ inches

The larva is often a brightly colored caterpillar, smooth and black with rows of many white, cream, or yellow-green *squarish markings*. The head, base of true legs and prolegs, and last segment are all *brilliant red*. Its long hairs are fine and light-colored. Its defensive posture is illustrated. Azalea Caterpillars feed heavily on azalea, often in colonies. They also eat apple and blueberry plants. Damage caused by these larvae is often evident in the Southeast around August and September, when the larvae are near maturity. The bare pupa spends the winter in the soil. The adult moth is called **Major Datana.**

YELLOW-NECKED CATERPILLAR To 2 inches

Though this larva always has *long yellow stripes*, its background color may be either orange or black. Its head is black, and as its name implies, its "neck" is yellow. Its hairs are sparse. It feeds upon a wide selection of host plants including crabapple, peach, cherry, maple, elm, walnut, and oak. The larvae feed in groups, all of them twitching simultaneously when disturbed. Notice its defensive posture. It is most common in the Northeast and Midwest but is found as far west as California. The adult Yellow-necked Caterpillar Moth is similar to the Major Datana, though paler in color. Both have wingspans of about 2 inches.

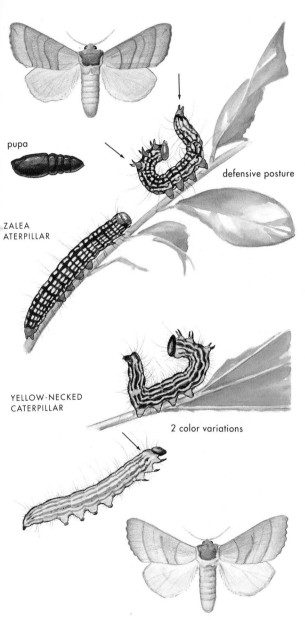

pupa

defensive posture

ZALEA
CATERPILLAR

YELLOW-NECKED
CATERPILLAR

2 color variations

91

Hairy

LARGE TOLYPE   To 1³/₄ inches

This caterpillar's body is *mottled* with white to gray splotches, a twiglike coloration that is good camouflage protection. It can be *hairy,* especially along the lower sides, where long hairs grow downward so that the prolegs are barely visible, enhancing its disguise. The larva grows slowly, feeding on willow, apple, ash, elm, oak, lilac, and other trees. The cocoon is tight, long, and pointed at both ends. It is attached to a twig on the food plant.

The adult female moth is about 2 inches across and has a hairy, white and gray body. The male is smaller. It is found in the eastern states west to Minnesota, Nebraska and Texas. The female lays her brown, dented eggs together in curved rows and covers them with hairs. The string of eggs looks much like a caterpillar.

Hairy With Tufts or "Pencils"

SPOTTED APATELODES   To 1¹/₂ inches

The larva is covered with yellow-white hair, which may look swept back. There are *9 black hair pencils,* or narrow clusters of long hair, along its back. The first two "pencils" and the last one are longer than the others (in the illustration, the larva's head is facing left). The body is white, and older larvae have broken vertical stripes of black on each segment and black markings along the back. This caterpillar feeds primarily on wild cherry but will eat other fruit trees, maples, and oaks. It is a placid caterpillar that feeds lightly on fresh leaves. The pupa overwinters in the soil. It is shiny and nearly black.

The adult moth, also called Wild Cherry Moth, has a wingspan of 1¹/₂ inches. It has brownish red patches on the lower inside edges of its gray forewings. It is found in the East, west to Wisconsin, Missouri, and Texas.

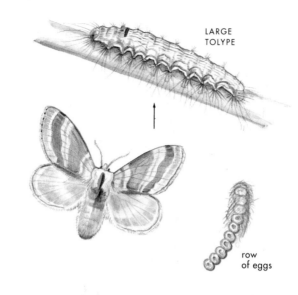

LARGE
TOLYPE

row
of eggs

SPOTTED
APATELODES

93

Hairy With Tufts or "Pencils"

AMERICAN DAGGER MOTH To 2 inches

The caterpillar is thickly covered with white, cream, or yellow hairs and has a *shiny black head.* Its most distinctive feature are its *5 long, narrow, black hair pencils,* four on the back and one near the rear. This larva feeds on many different trees; here it is shown on willow, but other hosts include maple, oak, basswood, and box-elder. The pupa overwinters inside a cocoon made of the larva's hair and leaves of the food plant.

The adult moth is usually grayish brown, with dark "daggers," or pointed zigzag stripes, on the forewings. It measures a little over 2 inches across and is common northeast of the Rockies.

BANDED TUSSOCK MOTH To 1¹/₃ inches

The caterpillar is densely covered with hairs that vary in color from gray or cream to yellow or olive green. It has both *black and white hair pencils* in front and rear. Two segments near the front each have two pairs of hair pencils, one black and one white, and there is a black pair and a white pair on rear segments. There is a *dark stripe* down its back. The head is large and shiny black, and there are yellow-orange markings with black stripes behind the first body segment. This caterpillar feeds on a variety of trees such as elm, ash, box-elder, cherry, and apple. Its small cocoon is hairy.

The Banded Tussock Moth, with a wingspan of up to 2 inches, is named for the dark bands on its translucent, buff-colored forewings. It is also called the Pale Tussock Moth for its light color. It is found in the East except in southern Florida and southern Texas.

AMERICAN DAGGER MOTH

cocoon

cocoon

BANDED TUSSOCK MOTH

Hairy With Tufts or "Pencils"

WHITE-MARKED TUSSOCK MOTH   To 1¼ inches

The mature caterpillar is hairy with *3 long black "pencils"* and *4 pairs of short, thick tufts of hair* that may be white or yellow. There are other sparse hairs along the sides and back. The body is yellow, with a black stripe with red spots along the back. The head is *yellow and orange.* This caterpillar feeds on many hosts, most commonly maple, horse-chestnut, birch, apple, sycamore, poplar, linden, and elm.

The adult male moth is generally small and brown. The female of this species is wingless. She lays her eggs in a frothy white mass on her empty cocoon; the eggs over-winter. This is an eastern species, found in deciduous and mixed woodland areas.

WESTERN TUSSOCK MOTH   To 1 inch

This caterpillar, like the White-Marked Tussock Moth larva, has *3 long black hair pencils:* 2 near the head and one at the rear. On its back there are *4 pairs of gray tufts.* The body is dark with many *orange-red tubercles* and *yellow markings* on the sides. It feeds on willow, hawthorn, oak, crabapple, and many other plants on the Pacific Coast. The light brown moth is similar to the White-Marked Tussock Moth.

MILKWEED TUSSOCK MOTH   To 1 inch

This caterpillar is thickly covered with tufts and pencils. Its body is brown, with *many black and white hair pencils* of different lengths along the front, rear, and sides. The middle segments also have *6 pairs of short, thick, yellow and black tufts.* The larva is found in late summer, feeding in colonies on milkweed. It may roll into a ball and fall to the ground when disturbed. A hairy cocoon protects the overwintering pupa.

The moth has a bright yellow abdomen with black spots. It lives in the eastern half of the United States.

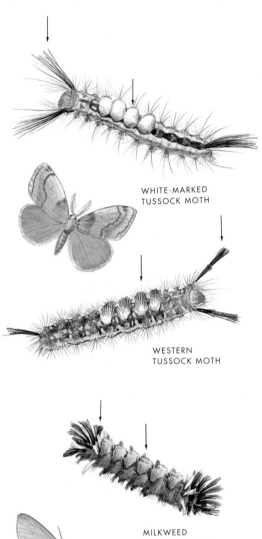

WHITE-MARKED
TUSSOCK MOTH

WESTERN
TUSSOCK MOTH

MILKWEED
TUSSOCK MOTH

Bristled

WOOLLY BEAR CATERPILLAR   To 2¹/₄ inches

Even those who know very little about caterpillars are familiar with the Woolly Bear,
which is commonly seen crossing roads in
the spring and fall. Its hairs are actually not
woolly to the touch but rather short, stiff,
and bristly. They cover the body very
thickly, a *red-brown band* of hairs in the
middle of the body and *black bands at both
ends.* Legend says the relative length of the
bands in the fall foretell the severity of the
coming winter. The color variations are in
fact due to many factors including the caterpillar's age; older larvae have more brown
hairs. Woolly Bears feed on dandelion, plantain, and many other low-growing weeds
and grasses. The caterpillar rolls up when
disturbed. Winter is passed by the caterpillar; the oval, hairy cocoon is made in spring
and summer by two generations per year.

The yellow-brown **Isabella Tiger Moth**
(wingspan 2 inches) is the adult of the
Woolly Bear. The female is darker than the
male. It is a very common moth throughout
the United States into Canada and Mexico.

SALT MARSH CATERPILLAR   To 2 inches

This caterpillar, a member of the same
family as the Woolly Bear, has a thick covering of short, bristly hairs. There are also
many long, whitish, *wispy* hairs, especially
near the rear and head. Its hair colors are
variable, from yellow and reddish orange to
dark brown. Its most distinguishing feature
is the *yellow streak* on the front of the head.
There may be light stripes along the sides of
the body. It feeds on many trees, garden
crops, and weeds, including apple, corn,
peas, and cord grass, which grows in salt
marshes. The caterpillar overwinters.

The adult Salt Marsh Moth has white forewings with black spots; the male's hindwings are yellow. The abdomen is yellow
with black spots. It is common throughout
the United States and is seen often in fall.

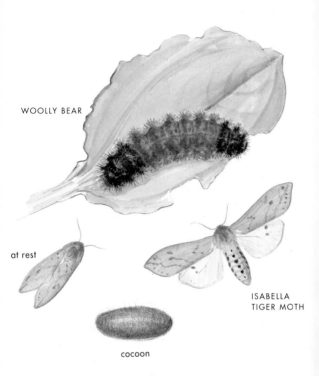

WOOLLY BEAR

at rest

ISABELLA
TIGER MOTH

cocoon

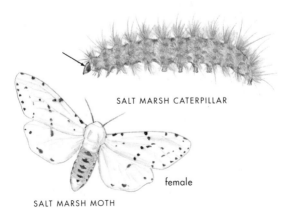

SALT MARSH CATERPILLAR

female

SALT MARSH MOTH

99

Bristled

GIANT LEOPARD MOTH   To 3 inches

This caterpillar can grow to be quite large. It
is thickly covered with short, thick, stiff,
black hairs. When threatened, the larva
rolls up into a ball, displaying the *vivid red
bands* between its body segments. It feeds
mostly on low-growing plants such as plantain, dandelion, violets, and the honeysuckle shown in the illustration, but it also
eats maples, and willows, and banana
leaves. The larva overwinters and spins its
cocoon in spring. It is a member of the Tiger
Moth family. If you rear this larva, be sure to
put some leaves of a food plant in its container, placed in a cool or cold (but not freezing) location over the winter. In the wild, the
overwintering larva hides in plant material,
and on warm days it might stir to nibble on
a few leaves. As the day cools, it resumes its
hibernation.

The adult moth is very striking, with its 3-
inch wingspan, white and black spotted
forewings, and metallic blue abdomen with
orange markings. It is eastern but more
common in the south to Texas.

VIRGIN TIGER MOTH   To 2 inches

The caterpillar is covered with *stiff, black,
bristly hairs*. It eats plantain, lettuce, clover,
cabbage, and other low plants. The larva
overwinters.

The adult Virgin Tiger Moth has attractive
cream and black patterns on its forewings,
orange hindwings with black patches, and a
wingspan of up to 3 inches. It is found from
Newfoundland to Florida and into the
Southwest.

GIANT
LEOPARD MOTH

VIRGIN TIGER MOTH

101

Branched Spines

RED ADMIRAL   To 1¹/₄ inches

Each segment of this common caterpillar's body has white and black *spines with pointed branches.* The spines can be orange or yellowish at the base. The segments are separated by *narrow white stripes.* The larva is usually dark (but can be lighter in color) with yellow side stripes and many white bumps with white hairs. The head is black, and the body can be white, yellow, green, brown or gray; young larvae are entirely black. Since this larva's colors vary so much, its food plant and shelter are important keys to its identification. The larva feeds in folded leaves of plants in the nettle family; here it is shown on stinging nettle. It overwinters as a pupa or adult. The chrysalis is not usually formed on the food plant, but rather hangs upside down from a nearby stem, twig, or firm upright plant structure.

The adult Red Admiral butterfly has flashy red diagonal bands crossing the forewings and red bands on the lower edges of the hindwings. Wings measure 2 inches across. It is easy to become familiar with this butterfly, because it is territorial and frequents the same location over periods of days or weeks. It will light on your head or clothing if you don't move quickly. The Red Admiral is a fast flier, often seen in sunny gardens, paths, and meadows. It feeds on many different garden flowers and basks in sunny spots with its wings spread. This warms the butterfly to help it fly. Fewer butterflies fly on cloudy days because there is less sunshine to warm their wings. It is found throughout North America, from the edge of the Arctic tundra south to Guatemala. It is also native to other parts of the world, including Hawaii, northern Africa, and Eurasia. Because it cannot survive cold winters, the northern parts of its range must be recolonized every spring by migrants.

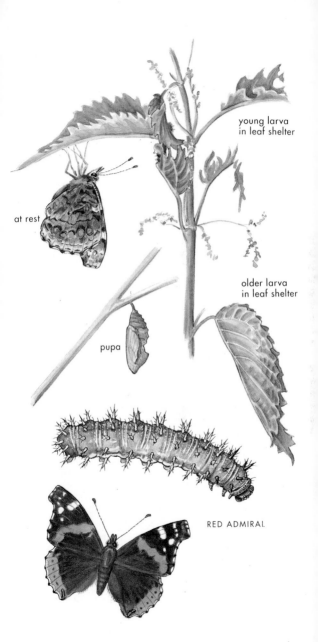

young larva
in leaf shelter

at rest

older larva
in leaf shelter

pupa

RED ADMIRAL

Branched Spines

PAINTED LADY   To 1½ inches

The body color of the larva is variable. It might be yellow-green, light purple, or grayish brown with many narrow yellow vertical stripes between segments, or it may be mostly black overall. Some of its many *branched spines* may be yellow, orange, or light-colored with black tips, but most are black. The larva always has a *dark line* along its back and a *yellow stripe* (usually broken) on each side. Its underside may be brown. The Painted Lady caterpillar makes a silk nest on top of the leaf where it feeds. It prefers to eat thistle but will eat many other related plants (composites) such as daisies and everlasting, also hollyhock and mallow. It overwinters as an adult in warm areas. Populations migrate north in spring and can become widespread, especially in wild, open areas.

The Painted Lady butterfly's remarkable ability to adapt to many environments has earned it another name, Cosmopolitan, as it is found in all but the most extreme environments throughout the world. Members of large populations migrate to less populated areas. It measures about 2 inches across.

AMERICAN PAINTED LADY   To 1¼ inches

The larva feeds on members of the daisy family, but prefers species of everlasting. It is a beautiful caterpillar, marked with contrasting *yellow-green stripes* and *red and white spots* on black bands. It also bears many *black branched spines.* The young larva is less colorful; it is small, black, and spiny. The caterpillar lives in a solitary nest made of silk and leaves and blossoms of the food plant. The chrysalis is brown or greenish with gold markings and may be formed inside the larva's nest. The adult is said to be quite tolerant of cold and can overwinter in northern states. It is found throughout the United States but it is more common in the East.

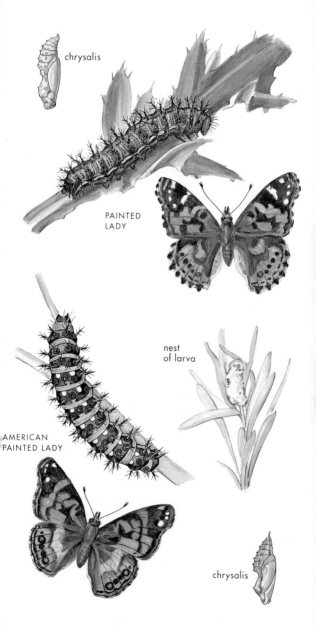

chrysalis

PAINTED
LADY

AMERICAN
PAINTED LADY

nest
of larva

chrysalis

105

Branched Spines

BUCKEYE   To 1¹/₄ inches

The spiny larva is usually black with 2 rows of orangish spots along the back and 2 rows of cream spots on the sides. The *branched spines* are *black,* with *blue bases* on the back spines and *orange bases* on the side spines. The prolegs are orange; the head is black with *orange* on the top and sides and has 2 short, black spines on top. The Buckeye caterpillar eats plantain, gerardia, toadflax, snapdragon, stonecrop, and loosestrife. Gardeners sometimes put these larvae in their gardens as a natural way to control weeds. The pupa is usually light cream with red-brown markings, but it can be nearly black. Larvae and adults overwinter.

The adult butterfly has 2 eyespots on each wing and orange-white diagonal bands on the forewings. It often basks in the sun, absorbing warmth for flying and feeding. It lives throughout the United States, except in the Pacific Northwest.

MOURNING CLOAK   To 2¹/₂ inches

This caterpillar is *black* with many *tiny white dots* and *black branched spines.* There are *8 red markings* on its back between the bases of spines. It is a long caterpillar when mature, having fed upon elm, willow, poplar, birch, hackberry, nettles, or wild rose. It can damage the foliage of shade trees. The pupa is cream and hangs upside down.

The adult is our largest butterfly in the tortoise shell family (see also page 108), with a wingspan of up to 3¹/₂ inches. The dark red-brown wings are edged with yellow bands. The adult may live nearly a year, spending the winter with wings folded, camouflaged by the underwings, in protected places such as in hollow trees or crevices in neglected buildings. It may be seen flying about on sunny late-winter days. The Mourning Cloak is common throughout most of North America and is native to Europe.

BUCKEYE

MOURNING CLOAK

Branched Spines

COMPTON TORTOISE SHELL   To 2 inches

The caterpillar varies in color. It may be light green with paler speckles and stripes, whitish with orange and black markings, or dark overall with light yellow markings as shown. There are always many *branched spines* (black or yellowish) and 2 branched spines on the dark head. Its food plant can be a clue to its identification: it feeds on white birch, aspen, and willows. The adult emerges from the pupa in midsummer and feeds on sap, nectar, and mud. It hibernates through the winter, then mates and lays eggs in spring.

The colors of the adult butterfly live up to its name, a mixture of black, brown, and orange. The angular wings can measure almost 3 inches across. It is a northern species, found across the northern half of the United States in deciduous woodlands.

MILBERT'S TORTOISE   To 1³/₄ inches
SHELL

The larva is *black* with greenish yellow sides. The body is covered with *whitish and orange dots* and many *black branched spines*. The head is black. Young larvae feed together in groups, then become more solitary as they mature. Pale green eggs are laid in clusters on nettle, the food plant. Use the food plant as a key to identification, because the colors of the larvae are variable. Winter is passed by the adult or the pupa. There may be three generations a year.

The adult butterfly, with a 1³/₄-inch wingspan, is smaller than other tortoise shells. It has the angled wings characteristic of tortoise shell butterflies. It sips nectar, sap, and mud. In midsummer, adults often migrate to higher altitudes to sip nectar from wildflowers, but return to lower areas for the winter. This northern butterfly is found through most of Canada and the northern parts of the United States, farther south in the West.

COMPTON
TORTOISE
SHELL

MILBERT'S
TORTOISE
SHELL

109

Branched Spines

QUESTION MARK   To 1¹/₂ inches

The spiny caterpillar is black with *white dots*. It has *2 orange stripes* on each side and *yellow or orange lines* along its back. Most of the spines are *orange*; the side spines are yellow. This larva and its near relatives have 2 black spiny clubs on the red-brown head. The Question Mark caterpillar feeds on elm, hackberry, nettle, basswood, and hops. Light green, cone-shaped eggs are laid singly, stacked or in short rows. The pupa hangs upside down from a plant part.

The adult butterfly is named for the silver question mark on the underside of the hindwing. The marking shows when the butterfly rests with its wings upright. The markings of the wings' undersides provide superb camouflage. With a wingspan of almost 3 inches it is the largest member of our anglewing butterflies, a group named for their angular wings. Adults overwinter. The Question Mark is common east of the Rocky Mountains into Mexico. It flies from early summer to fall.

HOP MERCHANT   To 1 inch

This caterpillar may be white, dark brown, black, red-brown, or green-brown. Darker larvae have light markings and crosslines. The spines on the head are black. Other body spines are whitish or yellow. The caterpillar feeds at night on the leaves of hops, elm, and nettle. The pale green eggs are stacked in a pile of up to 9 eggs.

The adult Hop Merchant is a common anglewing butterfly, widespread throughout the area east of the Rockies. It is also called Comma because the silver mark on the underside of its hindwing is comma shaped. It is smaller than its relative the Question Mark and is seen earlier in the year, when the butterfly emerges from hibernation in March. It sips sap and fruit but not flower nectar.

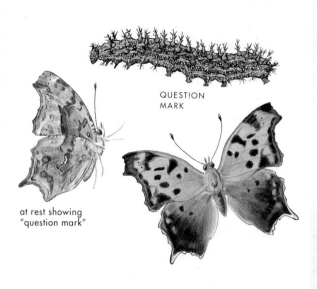

QUESTION
MARK

at rest showing
"question mark"

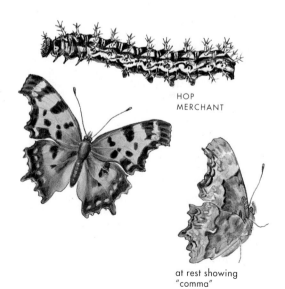

HOP
MERCHANT

at rest showing
"comma"

Branched Spines

BALTIMORE   To 1 inch

The caterpillar is marked with *narrow black bands* between *orange* segments. The head and branched spines are black. The young larva feeds in a silk nest on the food plant: turtlehead, false foxglove, or plantain. The chrysalis, which hangs upside down from a twig or stem, is white with black markings and orange dots. The partly grown larva overwinters.

This is one of the checkerspot butterflies, named for their checkered wing markings. The Baltimore is found in the East.

ANICIA CHECKERSPOT   To 1 inch

The caterpillar is *black* with branched spines that are mostly black, though the spines along the back and sides are *orange*. There may be *orange dashes* below the base of the spines on the sides. The white or yellow markings are variable. It feeds upon snapdragons and their relatives, such as Indian paintbrush, also plantain, snowberry, and honeysuckle. Immature larvae hibernate through the winter. The chrysalis is much like the Baltimore Checkerspot's.

The adult butterfly is black with white and red checkered markings. It populates fields and mountainsides in the West.

PEARL CRESCENT   To $3/4$ inch

The larva is *chocolate brown* with many small white dots, *white markings* along its back, and *cream lines* on each side. The branched spines are brown often with white tips. The spines on the cream lines have orange bases. The head is black with light markings. The adult female deposits egg clusters on asters, the caterpillars' food plant. The partially grown caterpillar overwinters. The chrysalis is gray, yellow, or brown.

The small butterfly, also called the Pearly Crescentspot, is common in meadows and open spaces east of the Rocky Mountains.

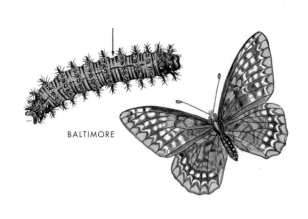

BALTIMORE

ANICIA
CHECKERSPOT

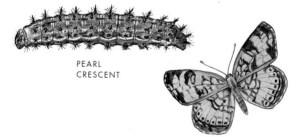

PEARL
CRESCENT

Branched Spines

GREECE SPANGLED FRITILLARY

GREAT SPANGLED FRITILLARY   **To 1³/₄ inches**

The larva feeds on the leaves of violets at night, as do most of the fritillaries. It is *black* and covered with branched spines that are *orange-red* at their bases and black at the tips. Winter is passed by the first-instar caterpillar. The chrysalis is mottled brown.

The large adult butterfly, with a wingspan of 3 inches, is a strong flier. It is orange with black spots. The silver "spangles" on the underside of the wings show when the butterfly is at rest, with its wings held upright. The Great Spangled Fritillary prefers damp meadows, where it sips nectar from many summer flowers, including thistles and black-eyed susan. It has a wide range, from the East, through the northern and central parts of the United States to California.

MEADOW FRITILLARY   **To ²/₃ inch**

The smallish caterpillar is purplish black with many brown, branched spines. It has dark, *V-shaped "arrowtip" markings* on its back and *dashes* on the sides of each segment. It feeds on violets. Eggs are white, the pupa brown with gold markings. The mature larva hibernates through the winter.

The small Meadow Fritillary butterfly frequents wet meadows and streamsides in the Northeast and north-central United States. It is sparse but present in the Northwest and into Canada. It flies low to the ground in zigzag patterns, feeding on low-growing flowers. The male gets the female's attention by hovering in front of her.

GREAT
SPANGLED
FRITILLARY

pupa

at rest

MEADOW
FRITILLARY

wings upright

pupa

115

Branched Spines

GULF FRITILLARY
To 1¹/₂ inches

The larva is black-brown with *reddish brown stripes* and *black branched spines;* the 2 spines on the head *curve backward.* It feeds on passion flower, which is toxic. This toxin stays in the larva, making it poisonous to predators. The yellow egg is long and narrow and has ridges. The chrysalis looks like a dried leaf; it is well camouflaged. None of its stages can survive freezing temperatures.

The 3-inch adult is not a true fritillary but one of the longwings, a group of butterflies whose forewings are often twice as long as they are wide. Most longwings are tropical; the Gulf Fritillary occurs along the Gulf of Mexico and much of the southern U.S., migrating north in summer. Its red-orange wings have black spots, and the undersides of the wings have silver spots like those of fritillaries. The Gulf Fritillary flies in sunny areas where wildflowers thrive.

ZEBRA BUTTERFLY
To 1¹/₂ inches

The caterpillar is *white* with *small black spots* and *long, black, branched spines.* There is a pair of spines on the head. It feeds on passion flowers found in southeastern United States. The chrysalis has a long, angular shape similar to that of the Gulf Fritillary, but the Zebra's mottled brown chrysalis is spiny. Neither can overwinter through freezing temperatures.

The adult is also called the Zebra Longwing for its long, narrow black wings with yellow stripes. Though it can be nearly 4 inches across, the Zebra is a weak flier. Adults roost together at night. It is mainly a species of tropical America, but its range extends north into Kansas and South Carolina. It can be seen in wood edges and thickets or hummocks where passion flowers grow. Like the Gulf Fritillary, the larva and adult of the Zebra are poisonous to birds and other predators because they retain the toxins of the larva's food plants.

GULF
FRITILLARY

chrysalis

ZEBRA BUTTERFLY

Branched Spines

NEVADA BUCK MOTH   To 3 inches

The caterpillar is whitish or greenish, with
small black markings, a *black line* along its
back, and a *red head.* It is covered with
branched spines that are *orange* along the
back and black elsewhere. If touched, the
spines can be painful or cause a rash, so
handle this caterpillar with care. Winter is
passed by the eggs, which are laid closely
together in a mass that circles a twig. The
larvae feed together on willow, oak, and
cherry, often causing damage to the trees.
At the end of the summer, the larva makes a
bare pupa in soil.

The day-flying adult moth measures up to
3 inches across and appears in the fall
around the time of deer-hunting season,
thus its common name, given around the
year 1900. It is a southwestern species; the
similar Buck Moth populates the East. The
buck moths, the Sheep Moth, and the Io
Moth belong to the same family as the giant
silkworm moths, such as the Luna, Cecropia, and Polyphemus (pages 50–59).

SHEEP MOTH   To 3 inches

The mature caterpillar is *dark brown,* with
red spots on its back and a red line along
each side. The short branched spines are
tan and black. The Sheep Moth larva's
spines are only mildly irritating to skin, but
it should be handled carefully. It feeds on
plants of the rose family that grow in pas-
tures and fields, often where sheep graze.
Eggs overwinter in masses around twigs.
Young black larvae emerge in spring.

The adult moth is generally pink with a
yellow streak and has variable black markings. It measures up to 3 inches across the
wings and flies low and strongly by day. It is
found from the Rocky Mountains west to the
Pacific Coast, from southwestern Canada to
southern California.

NEVADA
BUCK MOTH

SHEEP MOTH

Branched Spines

IO MOTH   To 3 inches

Young Io Moth larvae feed in groups, changing color twice as they grow to maturity. The young larva starts out orange with gray bristles, later adding red side stripes. Nearer to maturity, it is green with *red side stripes* and branched spines. These spines are irritating to skin, so handle the larva carefully. The larva rolls up and falls to the ground when disturbed. It feeds on cherry, maple, oak, beech, poplar, willow, corn, clover, roses, and many other trees and shrubs. The pupa overwinters in a cocoon made from leaf parts of the food plant. Eggs are laid in clusters and look like corn kernels.

The adult Io Moth measures 2–3 inches across. The male has yellow forewings; those of the larger female are browner. Both have black eyespots on their yellow hindwings. They are often attracted to lights on summer nights. The Io Moth is found in the eastern half of the United States.

SADDLEBACK CATERPILLAR   To 1 inch

This caterpillar gets its name from the *brown saddle markings* on its smooth green back. It has *stiff, stinging hairs* on fleshy filaments, giving it the appearance of having many branched spines. Two brown filaments extend from the front and two from the rear, and there are many shorter ones along each side. Brushing the skin against these spines can be quite painful. The Saddleback Caterpillar feeds on many plants, including apple, cherry, oak, asters, blueberry, rose, and corn. It is a small larva, just 1 inch long when mature.

The adult Saddleback Caterpillar Moth is less commonly seen than the larva. The female is larger than the male and has a wingspan of $1^1/_2$ inches. It is mainly an eastern species, ranging from Massachusetts to Florida and west to Missouri and Texas.

egg
side view

egg cluster

IO MOTH

female

male

SADDLEBACK
CATERPILLAR

121

Internal Feeders

In this group, eggs are laid on the host plant, and the larvae bore into the plant to feed on the juicy inner tissues. The larvae are smooth and mostly colorless—since they are hidden from predators, they have little need for protective spines or camouflage. Many of these caterpillars are considered pests because they damage desirable plants.

SQUASH VINE BORER   To 1$^1/_4$ inch

Find this caterpillar by looking for the damage it causes by feeding on squash, pumpkin, and melon plants: leaves may hang from their stems, or the entire vine might wilt. The mature white grublike larva emerges from a stem or fruit and burrows into the ground. It overwinters in the soil as a larva or pupa.

The adult moth is protected from predators by its resemblance to a wasp. This insect is considered a pest throughout its range, which includes most of North America except the Pacific Coast.

YUCCA SKIPPER   To 2$^1/_2$ inches

This caterpillar feeds in the leaves, stalks, and roots of species of yucca plants. The egg is laid on a leaf tip. The growing larva may tie a few leaves together to make a nest, later moving down through the leaf stalks to the root. The mature larva has little color, but the head is *black*. The larva overwinters.

The 3-inch adult skipper is seen in open grasslands of the Southeast and Southwest.

CODLING MOTH   To 1 inch

This small, smooth, white or pinkish caterpillar is the proverbial "worm" in the apple. The developing larvae bore into fruits or nuts to feed, causing serious damage to the fruits of apple, pear, and walnut trees. The waste, or frass, expelled from the entrance hole is an indication of the caterpillar's presence. Imported from Europe, the Codling Moth invades host plants wherever they grow in the United States and around the world.

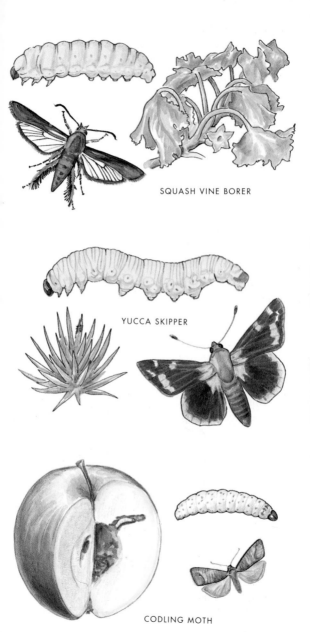

SQUASH VINE BORER

YUCCA SKIPPER

CODLING MOTH

123

Structure Building

SCALLOPED SACK-BEARER   To 1 inch

The newly hatched larva builds a zigzag net of silk threads under which it feeds. The more mature larva builds a cylindrical chamber from two leaves. This tubelike structure, open at both ends, serves as a shelter for the caterpillar. The larva may even cut its shelter from the leaf stems to make a portable sack in which it can travel safely about the food plant, a habit that earned it the name "sack-bearer." This small caterpillar is orange-brown when young; the mature caterpillar is greenish yellow and thick in the middle. Only the brownish front end is exposed while the larva is feeding in its sack. It may close one end of the sack for hibernation. The mature larva pupates in late spring. It feeds on oaks.

The small adult moth is called Scalloped Sack-bearer because of the deeply curved edges of the forewings. The egg is brownish and flattened and looks like a tiny bow tie. This moth is found in the East from New England to Florida, west to Iowa and Texas.

EVERGREEN BAGWORM   To 2 inches

Bagworm eggs hatch inside the protective case where they were laid. The young caterpillar comes out to feed and build its own individual case or bag, adding bits of plant material from the host plant. While inside the bag it can move about the food plant. It prefers conifers, juniper, arborvitae, and cedar but is also found on maples, oak, and sycamore. The larva feeds at the upright end of the bag, enlarging its bag as it grows up to 2 inches long. The mature larva pupates inside the bag. If it is a female, the wingless, legless adult never emerges. The male Evergreen Bagworm Moth (illustrated), with brown, transparent wings 1 inch across, emerges to mate. A mated female lays whitish eggs that overwinter inside her bag. Bagworms range from Massachusetts to Florida and west to Nebraska and Texas.

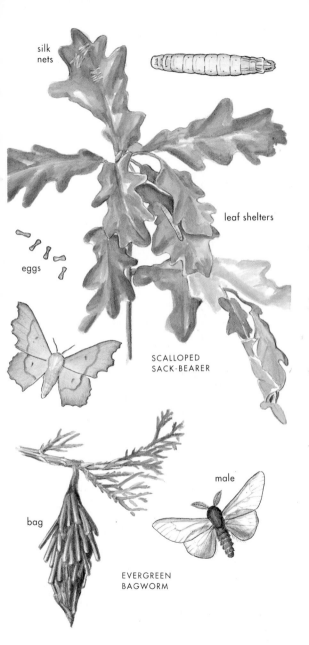

silk nets

leaf shelters

eggs

SCALLOPED SACK-BEARER

bag

male

EVERGREEN BAGWORM

125

Index